Evolutionary Process of a Steep Rocky Reservoir Bank in a Dynamic Mechanical Environment

To prevent the collapse of dangerous rock masses on steep rocky reservoir banks which can cause casualties and property loss, it is essential to design and conduct practical experiments to quantify the evolution processes of the reservoir banks and control such dangerous rock masses.

Using the Jianchuandong Dangerous Rock Mass project as a case study, this book generalizes the mechanical model of the project to show how improved equipment can be used to measure the mechanical state transition under the continuous action of axial pressure. It details a series of experiments to study the evolution of a severely steep rocky reservoir bank, which comprehensively consider the influence of hydraulic coupling, dry–wet cycles, axial pressure, and time-dependent effects. The results support a new method for determining the stability of dangerous rock masses on reservoir banks.

Features:

- Comprehensive application of methods involving on-site investigation, indoor experiments, theoretical derivation, and numerical simulation
- Combines engineering principles, real data, experimental methods and results
- Provides a complete research method for investigating hydrogeology failure processes

The book suits practitioners in hydropower engineering, engineering geology, and disaster protection.

Luqi Wang is a researcher in the School of Civil Engineering, Chongqing University, China.

Wengang Zhang is a Professor in the School of Civil Engineering, Chongqing University, China.

Challenges in Geotechnical and Rock Engineering

This series offers advanced level books focusing on state-of-the-art methods for handling problems across geotechnical engineering.

Chief Editor *Kok-Kwang Phoon, Singapore University of Technology and Design*

Assistant Editor *Dong-ming Zhang, Tongji University*

Evolutionary Process of a Steep Rocky Reservoir Bank in a Dynamic Mechanical Environment
Luqi Wang and Wengang Zhang

Evolutionary Process of a Steep Rocky Reservoir Bank in a Dynamic Mechanical Environment

Luqi Wang and Wengang Zhang

CRC Press
Taylor & Francis Group
Boca Raton London New York

CRC Press is an imprint of the
Taylor & Francis Group, an **informa** business

First edition published 2023
by CRC Press
6000 Broken Sound Parkway NW, Suite 300, Boca Raton, FL 33487–2742

and by CRC Press
4 Park Square, Milton Park, Abingdon, Oxon, OX14 4RN

CRC Press is an imprint of Taylor & Francis Group, LLC

ISBN: 978-1-032-38856-4 (hbk)
ISBN: 978-1-032-38858-8 (pbk)
ISBN: 978-1-003-34716-3 (ebk)

DOI: 10.1201/9781003347163

Typeset in Sabon
by Apex CoVantage, LLC

Contents

Challenges in geotechnical and rock engineering

Geotechnical and rock engineering have made significant strides in response to different challenges and opportunities that include measuring and understanding material/structural behaviours, handling special environmental conditions, performing complex numerical simulations, design and construction for low-carbon materials, novel structures, green construction and operation, life cycle and risk/reliability informed management, data-driven algorithms and AI-based decision making, digital twin and smart infrastructure, resilient engineering, climate change and sustainability, among others. These challenges are inter-related. Although some of the challenges are common to all industries, it is not meaningful to engage them in an abstract manner outside of the practice context. One example is data-centric geotechnics that takes a "data first" approach to decision making, but the data are actual "ugly" field observations (sparse, incomplete, spatially variable, corrupted, etc.) rather than ideal abstract numbers. Data-centric geotechnics should deploy digital technologies in cognizance of the context of geotechnical and rock engineering that include physics, empirical knowledge, experience, and engineering judgment. How decision making in geotechnical and rock engineering can be revolutionised through human machine collaboration is one grand challenge that epitomize the motivation for launching this book series. This book series presents exciting emerging solutions in geotechnical and rock engineering that are expected to transform practice and to meet fast evolving environmental/societal trends in the 21st century. It is a timely response to the changing technological, environmental, and societal landscape presented in the Institution of Civil Engineers (ICE) State of the Nation Report on "Digital Transformation" and the American Society of Civil Engineers (ASCE) "Future World Vision: Infrastructure Reimagined" paper.

Book series editors
Kok-Kwang Phoon
Dongming Zhang

Preface

Due to the periodic fluctuations of the water level, a degradation zone of up to 30 meters has been formed in the corresponding areas on both banks of the Three Gorges reservoir area. The toe of the dangerous rock mass on the reservoir bank is located in the degradation zone, which has experienced a complex mechanical environment for a long time and continues to deteriorate. This degradation trend promotes the overall failure process of a dangerous rock mass. Specifically, the base rock mass's evolution process involves the periodic rise and fall of the water level and the continuous deterioration of the rocky bank geomaterials. And this dynamic mechanical state makes it difficult to determine the stability of the dangerous rock mass. Compared with dangerous rock masses in other regions, the direct threat caused by the potential collapse of the rock mass is augmented by the potential swells, leading to more significant casualties and property losses. Therefore, it is essential to quantify the evolution process and minimize the dangerous rock mass's threat by designing and conducting practical experiments.

In this book, the Jianchuandong Dangerous Rock Mass (JDRM) project, located in the Three Gorges Reservoir area, was selected as the case history. Based on the field investigation, in-situ monitoring data, and laboratory test results, the mechanical model of the JDRM was generalized. The mechanical state transition under the continuous action of axial pressure was realized by improving the existing experimental equipment. Then, a series of experiments were carried out to study the evolution of the dangerous rock mass. These experiments comprehensively considered the influence of hydraulic coupling, dry–wet cycles, axial pressure, and time-dependent effect. Based on the testing results, a new method was proposed to determine the stability of dangerous rock masses on the reservoir bank. The main research contents and conclusions in this book can be summarized as follows:

(1) According to the tectonic evolution of Wuxia Gorge, the geological structure is complex, and the cliffs at the trailing edge are prone to collapse. As far, fluctuation of the reservoir water level has become an essential factor inducing the instability of the bank slope, and the hazardous zone

caused by the deterioration of the rocky bank continues to expand over time. For the mechanical environment of dangerous rock mass on the reservoir bank, a comprehensive survey system was constructed using various technologies and methods. This survey system can effectively detect the development of cracks and broken areas in cliffs, underwater sections, and the interior of the dangerous rock mass. The application of related methods can provide an important reference for the survey of similar submerged dangerous rock masses worldwide. Based on the filed survey, monitoring data, and indoor test, the mechanical model of JDRM was generalized. It can be found that the self-weight of the upper rock mass and the deterioration of the weak base of the rock mass caused by the fluctuation of reservoir water level were the main factors leading to the deformation and failure of the JDRM. This mechanical model was the basis of the experimental research in this book.

(2) An improved multi-scale mechanical experimental system was used to establish the complex mechanical environment of dangerous rock mass on the reservoir bank. Through this improved test equipment, the samples can complete the conversion in the dry state and the hydraulic coupling state while the axial pressure continues to act. The polarizing microscope, ultra-high-resolution scanning electron microscope, and energy dispersive spectrum were used to evaluate the microscopic mechanical properties of the selected samples qualitatively. Additionally, the quality of the sample was controlled to reduce the dispersion of the test data using computed tomography scanning and wave speed testing. The results show that the constituent minerals of samples were mainly calcite with few impurities. Meanwhile, the selected samples had few pores, and these pores were difficult to form the active seepage paths.

(3) The particle flow code (PFC), digital image correlation method, and acoustic emission technology were used to study the multi-scale evolution process under different mechanical states. As the rock mass changed from the dry state to the saturated state and then to the hydraulic coupling state, the influence of water on the mechanical properties of increased. Specifically, the participation of water reduced the bond contact, accelerated the deformation tendency between particles, and eventually led to increased plastic deformation and decreased peak strength. In the compression process under the hydraulic coupling state, the water would not be moved due to the compressed storage space; thus, the water pressure could continue to weaken the contact bond of the particles. This is the reason for the difference in mechanical properties between the saturated state and the hydraulic coupling state. According to the entire evolution process of rock mass investigated by the test, it can be found that the damage can magnify the effective stress experienced by the base rock mass. Moreover, incorporating the damage effects into the dangerous rock mass model created a nonlinear and accelerating deformation trend

that can accurately portray the failure of dangerous rock masses in the Three Gorges Reservoir area. Therefore, it is determined that the introduction of damage mechanics is reasonable and effective.

(4) A series of experiments were used to study the reservoir water's influence on the evolution of the submerged dangerous rock mass. According to the conventional dry–wet cycle test and numerical simulation, it can be found that under different mechanical states, the plastic characteristics and mechanical strength of rocks were quite different. And this difference was determined to be critical in accelerating the deterioration of rock mass. After that, the periodic transition of the mechanical state under different axial pressures were performed. The tests reveal that if the weight of the overlying rock mass were higher, the influence of mechanical transition on the base rock mass would be more significant. Furthermore, the evolution process considering the time-dependent effect was analyzed by the creep experiments under the transition of mechanical state. Compared with the creep experiments in a single state, the transition of the mechanical state continued to promote the deterioration of the rock mass, dramatically reduced the rock mass's peak strength, and accelerated the failure of the rock mass. Based on the field investigation and test results, the entire evolution of the dangerous rock mass on the reservoir bank can be divided into three stages: formation of the damage zone, progressive and cumulative deformation, and nonlinear accelerated failure. Firstly, the periodic rise and fall of the reservoir water level lead to the formation of a damaged zone. Then, the periodic transition of the mechanical state further promotes cumulative deformation, reduces the peak intensity of the base rock mass, and causes the dangerous rock mass to be in a critical state faster. Finally, when the axial pressure exceeds the long-term strength, the evolution would nonlinearly accelerate.

(5) After defining the mechanical state transition as the key factor in the degradation of rocky banks, a new method for determining the stability of the dangerous rock mass based on the changing reservoir water level was proposed. This method involved the generalization of a mechanical model, the derivation of the constitutive damage model based on laboratory results, iterative damage calculations, and the refinement of the safety determination based on the reservoir water level. Through this new method, the mechanical state of the dangerous rock mass can be continuously tracked, and the relationship between the changing water level and the potential failure can be quantified. After using this method to analyze JDRM, it can be found that the theoretical pressure data and the observed pressure data were consistent with one another, and both exhibited nonlinear deformation; this agreement shows that the new modeling technique was valid and robust. According to the parameters obtained by the prediction model, the dynamic collapse analysis of JDRM carried out in this book was similar to that of the Zengziyan collapse. It

not only proves the effectiveness of parameter selection but also expands this analysis method's capabilities. At the beginning of 2019, the treatment of JDRM was formally constructed. The simulation results indicate that anchoring the upper dangerous rock mass and reinforcement of the base rock mass can satisfactorily ensure the long-term stability of the JDRM.

The key points in this book can be concluded as follows:

(1) Based on the field investigation, indoor test, and in-situ monitoring data, the formation characteristics and the mechanical model of the dangerous rock mass on the reservoir bank were analyzed. Furthermore, the buckling failure mechanism of the submerged dangerous rock mass was first proposed.
(2) The influence factors were comprehensively considered, involving the axial pressure, mechanical state transition, and time-dependent effect. And then, the whole evolution process of the rocky bank was first reproduced indoors by the improved multi-scale hydraulic coupling test equipment. Moreover, the evolution model of the dangerous rock mass on the reservoir bank with nonlinear characteristics was obtained.
(3) By introducing the accumulated damage to the transition of mechanical state, a new method was put forward to determine the stability of dangerous rock masses on the reservoir bank. This method can quantify the effects of the reservoir water fluctuation and define the relationship between the damage and the macroscopic strength.

Keywords: Evolution process; Steep rocky reservoir bank; Dynamic Mechanical Environment; Numerical simulation; Damage evolution

Chapter 1

Introduction

1.1 Research motivation and significance

Rock mass collapse is a common geological disaster in mountainous areas. Among these collapse types, the steep dangerous rock mass is characterized by a sizeable height-diameter ratio, a wide stress distribution area, a high frequency of occurrence in nature, and the potential for rapid failure and widespread destruction (Yin et al., 2011; Xing et al., 2016; Feng et al., 2014a, 2014b; Ferrero et al., 2016; Zheng et al., 2018). As for the steep rocky reservoir bank, the direct threat caused by the potential collapse of the rock mass is augmented by the potential for swells, which can also cause significant casualties and property losses (Yin et al., 2015, 2016; Zhao et al., 2018; Xu et al., 2022). Therefore, it is essential to quantify the evolution process and control the threat of these submerged steep dangerous rock masses.

The periodic fluctuations of reservoir water level create a degradation zone along the bank slope (Yin et al., 2016; Tang et al., 2019; Zhang et al., 2022a, 2022b). Since the foot of the rocky banks is located in this degradation zone, it continues to deteriorate during the reservoir water fluctuation; this cyclical damage significantly impacts its overall instability (Zhou et al., 2016; Deng et al., 2019). Notably, the fluctuation of the reservoir water level not only reduces the macroscopic strength of the base rock mass but also imparts further damage to the base rock mass. The combination of these two factors affects the dangerous rock mass mechanically and environmentally, making it difficult to accurately evaluate these conditions using conventional analytical methods (Hu et al., 2018; He et al., 2019a, 2019b).

While the existing methods can judge the stability of a dangerous rock mass that is not submerged (Hungr et al., 2014; Ferrari et al., 2016; Huang et al., 2016a; Keneti and Sainsbury, 2018), these methods do not reflect the dynamic processes experienced by a dangerous rock mass on the reservoir bank. To solve this problem, a series of experiments were performed through improved experimental equipment.

In this book, the Jianchuandong Dangerous Rock Mass (JDRM), located in the Three Gorges Reservoir area, was selected as a case study. Based on the

DOI: 10.1201/9781003347163-1

field investigation, on-site monitoring data, and laboratory test results, the mechanical model of the JDRM was generalized. The mechanical state transition under the continuous action of axial pressure was realized by improving the existing experimental equipment. Through this improved equipment, a series of experiments were carried out to study the evolution of the dangerous rock mass. The influence factors such as hydraulic coupling, dry–wet cycles, axial stress, and the time-dependent effect were comprehensively considered in the research process. On the grounds of the test results, a new method was proposed to determine the stability of dangerous rock masses on the reservoir bank. After including the accumulated damage during the mechanical state transition, the new method could not only quantify the effects of the periodic fluctuation of the reservoir water level but also define the relationship between the damage and the macroscopic strength. After extracting the parameters from the critical state, the dynamic failure process and the stability trend after the treatment were determined. The related calculation methods can provide significant references for the analysis of dangerous rock masses on the reservoir bank.

Based on the existing research background, a comprehensive view of the present research is presented in this section, including the analysis method of the dangerous rock mass, the influence of reservoir water fluctuation on the rocky bank, and the rock damage mechanics.

1.2 Overviews on present research

1.2.1 The analysis method of dangerous rock mass

1.2.1.1 The stability of dangerous rock mass

There are generally four methods for evaluating the stability of hazardous rock masses: qualitative methods, quantitative methods, numerical and physical simulation methods, and uncertainty analysis methods. These methods will be briefly described in this section.

The qualitative evaluation methods mainly include the engineering geological analogy method and the stereographic projection method. Specifically, the various influential factors can be considered comprehensively by the engineering geological analogy method (Zhang et al., 2022c, 2022d). And then the stability and development trend of dangerous rock masses can be predicted quickly (Kilburn and Petley, 2003; Yu et al., 2014; Sun et al., 2014; Gao, 2015; Zhang et al., 2021). In addition, the stereographic projection method is widely used in the preliminary judgment of unstable rock masses (Tomas et al., 2012; Jia et al., 2013; Zhou et al., 2017a). However, due to the difference in dangerous rock masses' geological conditions, the accuracy of qualitative evaluation methods depends on the experience of the researchers.

The limit equilibrium analysis is the primary method to analyze the dangerous rock mass quantitatively. This method can investigate the stability of rock mass that tends to slide down under the influence of gravity. Translational or rotational movement is considered with an assumed or specific potential slip surface under the rock mass (Abramson et al., 2002; Zhu et al., 2003; Asadollahi and Tonon, 2012). The limit equilibrium method can be feasible and straightforward to obtain precise results. After continuous revision and improvement, this is the most commonly used method to determine the mechanical state of dangerous rock masses (Deng et al., 2016; Nilsen, 2017; Tonon, 2020).

Based on the principle of similarity and field survey results, physical model experiments are used to establish a simulation model (Adhikary and Dyskin, 2007). Under certain conditions, the physical model experiments can predict or reproduce the evolution of dangerous rock masses. For instance, Wang et al. (1991) studied the Lianziya dangerous rock mass in the Three Gorges Reservoir area through model experiments. Feng et al. (2012) investigated the failure mechanism of Jiweishan landslide through the geotechnical centrifugal model experiment. Their results reproduced the driving-blocks-and-key-blocks mode of apparent dip. He (2015) analyzed the evolution process of the Zengziyan collapse through the centrifuge experiment. However, the cost of physical model experiments is generally high, and the experimental conditions are difficult to control.

The numerical simulation can obtain the dangerous rock mass's stress or strain distribution and reproduce the dynamic failure. When performing numerical calculations, the finite element method (Clough and Woodward, 1967; Griffiths and Lane, 1999; Bendezu et al., 2017), the discrete element method (Cundall and Strack, 1979), and the discontinuous deformation analysis method (Shi, 1988) are commonly used. These numerical simulation methods have different operating mechanisms and applicable scopes. For example, the finite element method is widely applied to analyze continuous deformation (Jiang and Magnan, 1997; Lim et al., 2017; Li et al., 2019). Notably, it is important to determine the geological model, mechanical model, and mechanical parameters during the numerical simulation. And these settings are the key factors to ensure the simulation results' reliability and validity (Tang, 1997; Stead et al., 2006; Stead and Wolter, 2015; Sarfaraz and Amini, 2020).

Since there are many factors affecting the stability of dangerous rock mass, mathematical methods are adapted to determine the stability quantitatively or semi-quantitatively, such as dissipation theory (Qin et al., 2019), chaos theory (Gao, 2013), stochastic theory (Mitchell and Hungr, 2017), fuzzy theory (Daftaribesheli et al., 2011), grey system theory (Liu and Cao, 2015), catastrophe theory (Xia et al., 2015), and so on. These theoretical methods can help us better understand the evolution process to a certain extent and provide a reference for the stability analysis of dangerous rock mass.

1.2.1.2 Analysis of dynamic collapse of dangerous rock mass

Under the action of natural or human factors, the dangerous rock mass is separated from the bedrock and moves downhill in the form of a rolling stone, and the collapse will cause severe damage to the buildings and residents at the slope surface and the foot of the slope (Liu, 2014). The kinematic formulas (Azzoni et al., 1995), physical mechanics tests (McDougall and Hungr, 2004), and numerical simulations (Thompson et al., 2009) are used to analyze the dynamic collapse characteristics of dangerous rock masses. Factors that affect the movement characteristics include the mechanical properties of the falling rock and the actual conditions of the slip path. And the typical movement forms of rock collapse can be divided into sliding, straight falling, rolling, and jumping (Iverson et al., 1997; Sonnekus and Smith, 2018).

In terms of the derivation of kinematic formulas, Reznichenko et al. (2011) outlined the process of rock avalanche, which indicated that supraglacial deposits of rock avalanche debris had distinct sedimentological and thermal properties. Shen et al. (2012) divided the movement process into three stages: free-fall, bouncing, and rolling. Cheng and Su (2014) studied the movement characteristics and hazard range parameters of collapsed stones on slopes according to the site measurement and statistical analysis of the collapse induced by the Wenchuan earthquake. In the case of the Mabian landslide, Wang et al. (2020a) added a reduction factor lambda to the Scheidegger formula and achieved a modified formula that can determine the velocity of consequent landslides accurately.

In terms of the physical mechanic test, Ya et al. (1996) analyzed the collapse of Lianziya dangerous rock mass through the geo-mechanical model experiments. Huang et al. (2007) studied the effects of slope features on the rock-stopping position, the moving time, and the moving track of different rock block shapes. Bowman et al. (2012) investigated whether test conditions that produce dynamic fragmentation can lead to more significant runout or spreading of physical model rock avalanches. Wang et al. (2016a) conducted a series of experimental tests to document the propagation and deposit features of rock avalanches under the effect of 3D complex topography. Chen and Orense (2020) conducted laboratory experiments to release dry rigid blocks on an inclined chute to investigate the mechanisms involved in the downslope motions of granular particles.

In terms of numerical simulation, Xing et al. (2017) used DAN3D to simulate the post-failure behavior of the rock avalanche triggered by the Wenchuan earthquake in the Wenjia valley. Through trial and error, Li et al. (2017a) found that a combination of the frictional model and the Voellmy model can provide the best performance in simulating the rock avalanche triggered by the Lushan earthquake. Scaringi et al. (2018) adapted various models (PFC, MatDEM, MassMov2D, Massflow) to reproduce the Xinmo landslide and simulate the kinematics and runout of the potentially unstable mass. Liu et al. (2019) studied the dynamic collapse based on the block discrete element method. Bi et al. (2019) used discrete element methods to study the effects of the configuration of a baffle-avalanche wall system on rock avalanches in Tibet Zhangmu.

1.2.1.3 Research on the steep dangerous rock mass

The steep dangerous rock masses are widely distributed in southwest mountainous areas in China, such as Sichuan Province, Chongqing City, Yunnan Province, Guizhou Province, Hunan Province, and Hubei Province. Due to its unique development environment and structural characteristics, the steep dangerous rock mass has a wide distribution, a high frequency of occurrence, and a broad impact range, which poses a severe threat to life safety and urban construction (He, 2015). For steep dangerous rock mass, researchers have studied the influence factors, stability, and collapse mechanism from the aspects of geomorphology, lithology, and joint fissures.

Powell (1875) found that the steep rock masses, located in the Colorado River and its tributaries, were formed by long-term flow erosion and weathering. Terzaghi (1950) cited the progressive failure of the base area as the result of the Pulverhörnd collapse located in the Austrian Alps. Poisel et al. (1991, 2005, 2009) conducted considerable research on the collapse mechanism of steep hard rock located on soft rock. They found that the underlying soft rock's extrusion would cause the upper hard rock's deformation and further promote the rock mass to undergo translational sliding or rotational sliding. Nichol et al. (2002) used the controlled joints as the vital factor for the dumping failure of large-scale thick-layer limestone. Dussauge-Peisser et al. (2002) pointed out that the cliffs formed by limestone often develop near-vertical joints and differential weathering. This phenomenon caused the dangerous rock mass to be prone to dumping failure. Frayssines and Hantz (2006) carried out statistical tests to study the relationship between rock falls and daily rainfall, freeze—thaw cycles, or earthquakes. Based on the detailed consideration of the characteristics of the rock mass, Hu and Zhao (2006) adopted the related pivotal factors to demarcate different structure styles of rock mass in the slope of the red bed. Zhang et al. (2007a) used FLAC3D to evaluate and predict the stability of a steep dangerous rock mass that was likely to collapse. Li (2012) studied the bending and breaking of steep dangerous rock mass from the perspective of quality evaluation of rock. He et al. (2019a, 2019b) investigated the kinematic characteristics of Zengziyan collapse by image analytic method and then put forward the related collapse mode. Huang et al. (2016b) improved the judgment formula for the collapse of the dangerous rock mass on the reservoir bank. Wang et al. (2016b) considered the different mechanical states of dangerous rock mass during the fluctuation of reservoir water level. Wang et al. (2020b) adapted the finite element method to analyze the treatment of steep dangerous rock mass on the reservoir bank.

1.2.2 Influence of reservoir water fluctuation on the rocky bank

For the rocky bank, the impact of reservoir water fluctuation is complicated. This influence can be considered from three aspects: dry–wet cycles, hydraulic coupling, and time-dependent effect.

1.2.2.1 The dry–wet cycles

The dry–wet cycle is a dynamic process involving saturation, drying, and test methods after the cycles. Specifically, in the saturation process, it is necessary to consider a series of factors such as the pH level of the solution, the immersion time and method, and the pressure used for the saturation. In the drying process, it is required to determine the drying method and drying time. After the dry–wet cycle, it is essential to confirm the test method to obtain the rock samples' mechanical parameters. These factors have an important effect on the macro-mechanical strength, mineral composition changes, and microstructure of the rock under the action of the dry–wet cycles. Therefore, the realization process of the dry–wet cycle is significant for the test results.

Researchers have studied the changes in mechanical parameters of rock mass under dry–wet cycles, such as shale (Liu et al., 2013), tuff (Özbek, 2014), mudstone (Torres-Suarez et al., 2014), sandstone (Sumner and Loubser, 2008), siltstone (Jeng et al., 2000), altered granite (Chen et al., 2019), and limestone (Huang et al., 2016a). The test methods included wave velocity test and uniaxial compression test (Yao et al., 2013), conventional triaxial compression test (Yao et al., 2010), split test (Zhu et al., 2012), loading–unloading test (Torres-Suarez et al., 2014), direct shear test (Dai et al., 2022), rheological tests (Wang, 2014), and so on. Also, Scanning Electron Microscope (SEM; Coombes and Naylor, 2012) and Computed Tomography (CT; Wang, 2016) were used to analyze the micro-scale deterioration trend under dry–wet cycles.

The effects of dry–wet cycles are different for rocks with different lithologies or different mineral compositions (Xue and Zhang, 2011). For example, with the number of dry–wet cycles increasing, the change rate of the red sandstone's strength gradually decreased (Liu et al., 2016). However, the acceleration rate of limestone still didn't converge after 60 dry–wet cycles (Yang, 2011).

From the perspective of the rock mass's microstructure, during the dry–wet cycles, the cement inside the rock mass dissolves; the closed pores are converted into open ones; the microcracks develop on the particle contact surface caused by the frequently expand and compression. These phenomena are more evident in soft rocks that can quickly disintegrate in water (Zhang et al., 2015, 2018; Zeng et al., 2017; Liu et al., 2018b).

1.2.2.2   The hydraulic coupling

Many researchers simplify the mechanical state of reservoir banks into two types: dry state and saturated state (Wang, 2014; Liu et al., 2016; Huang et al., 2016a; Deng et al., 2019). However, the mechanical environment of the reservoir bank is more complex. The rocky bank not only bears the self-weight of the overlying rock mass but also is affected by the reservoir water fluctuation. When the water level rises, the mechanical environment of the reservoir bank should be defined as the hydraulic coupling state. Researchers have conducted in-depth research on the failure process of rock mass under hydraulic coupling (Rutqvist and Stephansson, 2003; Neuman, 2005).

In terms of laboratory tests, most researchers use the complete stress–strain process experiment (CSSPE) to study the evolution of permeability during rock damage and fracture. Chen and Lin (2004) studied the stress–strain–electric resistance experiments for diabase, limestone, and marble containing NaCl solution during the whole process of uniaxial compression. Zhang et al. (2007b) conducted laboratory experiments on massive concrete blocks with randomly distributed fractures and rock core samples. And these experiments were used to investigate fluid flow and permeability variations under uniaxial, biaxial, and triaxial complete stress–strain processes. Cheng et al. (2017) presented some detailed observations and analyses for a better understanding of the failure mechanism and seepage behavior of the samples under different confinements and water pressure.

In terms of theoretical research, there are currently two main methods. One approach is to establish the relationship model between stress–strain and permeability based on the test results of rock permeability in the CSSPE process; the other method is to create the permeability model during rock failure by using the damage mechanics theory. Bai and Elsworth (1994) proposed a model to describe the sensitivity of hydraulic conductivity to effective stresses through Hertzian contact of spherical grains. A three-dimensional (3D) finite element model that considered the coupled effects of seepage, damage, and stress was used to investigate the hydromechanical response of rock samples at a laboratory scale (Li et al., 2012a). Li et al. (2017a) presented a discussion of the state-of-the-art use of discrete fracture networks (DFNs) for modeling geometrical characteristics, geo-mechanical evolution, and hydro-mechanical behavior of natural fracture networks in rock.

Some researchers have also adopted different numerical simulation methods to study the hydraulic coupling characteristics of rock damage. For example, a finite element code (F-RFPA) coupled with the flow, stress, and damage analyses were used to study the influence of heterogeneity of mechanical properties on hydraulic fracturing in permeable rocks (Yang et al., 2004). Damjanac and Cundall (2016) applied the Distinct Element Method (DEM) to the simulation of hydraulic fracturing. Yan et al. (2016) presented a coupled hydro-mechanical model based on the combined finite discrete element method to simulate hydraulic fracturing in complex fracture geometries.

1.2.2.3 The time-dependent effect

During the reservoir water fluctuation, the reservoir bank would be under a mechanical state for a specified period. Therefore, after the mechanical state transition, there is sufficient time for the reservoir bank to complete the stress balance, involving the time-dependent effect on the rock mass. The creep experiments and related constitutive models are generally used to study the time-dependent effect on the rock (Cai et al., 2006). According to the creep tests, the relationship between the strain and time can be described by the constitutive model (Wang et al., 2016c; Wu et al., 2017). The current creep models can be divided into three types: empirical models (Griggs, 1939; He et al., 2007; Yang et al., 2013a), composite element models (Nomikos et al., 2011; Jiang et al., 2013; Yang et al., 2013b; Wang et al., 2018), and theoretical models that are based on internal time theory, fracture mechanics, and damage mechanics (Valanis, 1971; Lux and Hou, 2000; Adaehi et al., 2005; Zhou et al., 2011).

Significant research and progress have been made toward understanding the creep behavior of rock. For example, Tomanovic (2006) carried out laboratory tests on marl creep to formulate the creep model of time-dependent deformations of soft rocks. Sosio et al. (2008) used the landslide front velocities for back analyses and calibration of the creep parameters, together with the final shape and thickness of the deposit, and its developing extension. Zhou et al. (2011) proposed a new creep constitutive model by replacing a Newtonian dashpot in the classical Nishihara model with the fractional derivative Abel dashpot, and this model was based on the time-driven fractional derivative. Xu et al. (2012) carried out the triaxial rheological tests under seepage pressure by using rock servo-controlled triaxial creep testing equipment. They also studied the rock creep properties influenced by seepage–stress coupling and analyzed the variations of seepage rate with time in the complete creep processes of rock. Nedjar and Le (2013) developed a three-dimensional phenomenological model to describe the long-term creep of gypsum rock materials. Yang et al. (2013c) examined a rheological model to analyze the slope's creep behavior, predict the long-term stability, and guide appropriate measures to be taken at suitable times to increase the factor of safety to ensure stability. Zhao et al. (2017) studied the extensive nonlinear rheological characteristics of hard rock under cyclic incremental loading and unloading. Wang et al. (2018) performed creep tests using representative landslide sandstone in the Three Gorges Reservoir and investigated the sandstone creep behaviors under the coupling effects of seepage pressure and stress. Guo et al. (2020) adapted the DEM method to study the time-dependent behavior of hard sandstone.

1.2.3 State-of-the-art research of rock damage mechanics

There are usually faults and weak interlayers inside the rock mass so the rock mass can be defined as an anisotropic, discontinuous mechanical material with initial damage. The damage mechanics of rock mass is mainly used to study the whole evolution process of microcracks involving initiation, expansion, and penetration. The damage constitutive model can be established to analyze the damage degree of rock and evaluate the stability of rocky slopes (Cai et al., 2006).

Due to the fluctuation of the reservoir water level, the rocky reservoir bank has been in progressive destruction. Through the damage mechanics, the relationship between the mechanical state and the macroscopic strength can be obtained, which can help us better understand the evolution process of the reservoir bank. Therefore, it is vital to analyze the research actuality of rock damage mechanics.

Kachanov (1958) first proposed the concepts of damage mechanics and damage factors. Then Dougill and Al (1976) introduced damage mechanics into geotechnical engineering, and Dragon and Mroz (1979) proposed the concept of fracture surface for continuous damage mechanics of rock and soil. After that, Krajcinovic and Fonseka (1981), Krajcinovic (1984), Kachanov (1982), and Kemeny and Cook (1986) continued to promote the application of damage mechanics. At present, the application fields of damage mechanics involve static process, dynamic process, elastic material, elastoplastic material, intact rocks, and incomplete rocks with fractures. However, due to the complexity and uncertainty of rock, as well as the limitations of theoretical methods and test conditions, the damage process of rock mass remains to be further studied.

Many researchers have analyzed and verified the damage constitutive model of rock through numerical simulation. Bargellini et al. (2006) used the DEM to simulate the anisotropic damage mechanism of rock caused by microcracks. By analyzing the attenuation of sound wave propagation in brittle rock samples during the impact test, Wang et al. (2008) introduced the damage evolution to the finite element code LS-DYNA. Zuo et al. (2010) proposed a damage model of brittle materials under dynamic load through numerical simulation. Zhao et al. (2012) adapted the particle hydrodynamic analysis model to study the shear process of rock fractures. Chen et al. (2012) studied the micromechanical damage of brittle rock under multifield coupling (temperature, seepage, and stress). Ghazvinian et al. (2014) proposed a numerical calculation method based on DEM. This method can simulate the development of crack damage in brittle rocks by constructing a three-dimensional random polygonal structure. Liu et al. (2019) investigated the influence of rock damage on rock fragmentation and cutting performance by using PFC2D. Wei et al. (2020) input the inversed damage as the numerical simulation's initial conditions to predict the future damage and failure of rock.

Experimental data has always been the main basis for constructing damage constitutive models. Yuan and Harrison (2006) studied the damage characteristics of low-porosity crystalline rocks and high-porosity sedimentary rocks during brittle fractures. Shao et al. (2006) investigated the anisotropic damage and creep model of brittle rock and proposed a constitutive model based on micromechanics to study the damage of anisotropic brittle rock. Shao et al. (2009) examined the effect of blasting on rock damage. According to the relationship between critical fracture characteristics and damage variables, they proposed a damage evolution equation under a statistical damage mechanics network by considering the effects of loading rate and initial damage in rock. Based on energy theory and damage mechanics theory, Zhou and Zhu (2010) established an elastoplastic damage constitutive model with dual yield surfaces and verified the rationality of the proposed model by comparing the numerical prediction of the triaxial compression test with the experimental data. Li et al. (2012b) proposed a statistical damage constitutive model that can reflect the strain-softening of the rock mass and studied the validity of the model with examples. Erarslan and Williams (2012a) investigated the relationship between deterioration damage of rock mass and the shear strength of structural planes. Erarslan and Williams (2012b) studied the mixed failure model of rock under static and cyclic loading. Through scanning electron microscope images, they found that the damage of tuff cyclic loading was mainly caused by intergranular and transgranular fractures. Using a microscale damage mechanics model and a new geometric model for crack propagation, Tang and Zhu (2020) derived the macroscopic strain function of the geometric parameters of damages under specified loads.

Besides, many researchers have applied the damage mechanics to actual working conditions. Diederichs et al. (2004) studied the damage mechanism of the crack propagation of brittle rock during tunnel excavation. Zhao et al. (2004) proposed a damage mechanics method to predict the damage distribution of jointed rock masses during excavation based on continuous damage mechanics. After considering the response factors of rock mass under tensile and compressive loads, Molladavoodi and Mortazavi (2011) used the damage tensor to develop a new model that was closer to the real working conditions. Doan and d'Hour (2012) discussed the effect of initial damage of rock mass on the distribution of fault fracture zone. Mortazavi and Molladavoodi (2012) applied the damage constitutive model to the excavated roadway and verified the model in combination with on-site monitoring data. Based on the statistical parameters of the fracture network of rock mass, Wu et al. (2020) examined the elastic stress–strain relationship of the fractured rock mass. And then, they compared the monitoring data from a tunnel excavation with the theoretical results.

1.3 Shortcomings and prospects in research

So far, many researchers have made contributions to the analysis of the dangerous rock mass and have achieved abundant results and enormous progress. However, due to the complexity of the mechanical environment in the reservoir area, the evolution of dangerous rock mass on the reservoir bank is still an important issue to study further. The main shortcomings and the corresponding development prospects are summarized as follows:

(1) Regarding the failure mode of dangerous rock masses, most previous studies have considered dangerous rock masses as a whole, classified them as a slip or dump failure, and then determined the static stability based on their macro-strength (Hungr and Evans, 2004; Hungr et al., 2014; Huang et al., 2016a). These analysis methods can effectively determine whether a dangerous rock mass is in a critical state, especially for an unstable rock mass in a single mechanical state. However, due to the periodic change in the mechanical environment, the dynamic evolution related to the changing reservoir water should be considered. Under this unique mechanical state, the degradation zone is formed on the reservoir bank. And this differential weathering gradually becomes more obvious as the number of hydrological cycles increases, leading to the evolution characteristics of progressive deformation and sudden failure. Furthermore, when studying the dynamic collapse through numerical simulation, the determination of corresponding parameters lacks the theory's verification.

(2) Due to the limitations of conventional testing equipment, experiments conducted on the rocky reservoir bank have focused on the degradation of the rock mass caused by a single factor. Individually, the effects of dry–wet cycles on the mechanical properties of the rock mass were considered (Wang et al., 2017, 2020b; Deng et al., 2019). The hydraulic coupling tests were used to study the evolution of the reservoir bank (Wang et al., 2020c). And the deformation of the rock mass caused by the time-dependent effect was analyzed using creep experiments (Bozzano et al., 2012; Yang et al., 2013b; Wang et al., 2018). However, there is still a significant gap in achieving actual complex working conditions, as previous studies have not considered the combined action of mechanical state transition under the time-dependent effect. This knowledge gap is likely to lead to misjudgment of the real-time evolution process of the rock mass.

(3) The damage evolution of the reservoir bank caused by water fluctuation is complicated. It is required to consider both the time-dependent effect and the macroscopic strength degradation when analyzing this dynamic evolution process. Therefore, the existing damage mechanics methods should be improved by combining with on-site investigations, laboratory tests, and monitoring data. On the one hand, it is vital to introduce the reservoir water fluctuation into the damage mechanics formula; on the

other hand, it is essential to link the damage mechanics with the deterioration of macroscopic parameters. Meanwhile, damage mechanics' relevant settings need to be further simplified to make the research method easy to popularize.

The formation characteristics and evolution process of steep dangerous rock mass on the reservoir bank are affected by the self-weight and the reservoir water fluctuation. Therefore, the influence factors should be comprehensively considered by effective experiments, involving the axial pressure, mechanical state transition, and time-dependent effect. Moreover, it is essential to put forward a new method to quantify the effects of the reservoir water fluctuation and define the relationship between the damage and the macroscopic strength. And these are the key contents of this article.

1.4   Research method

1.4.1   Research contents

The main research content in this book can be summarized as follows:

(1) Through the analysis of the geological conditions of Wuxia Gorge, the fluctuation of the reservoir water level was considered an important factor inducing the instability of the rocky bank. For the specific geological environment of the high and steep reservoir bank, a promoted field survey system was proposed. Based on the filed survey, on-site monitoring data, and indoor test, the mechanical model of dangerous rock mass on the reservoir bank was generalized. Then, the existing equipment was improved to construct an experimental environment that can realize the mechanical state transition under continuous axial pressure. Meanwhile, a variety of techniques were used to determine the microscopic mechanical properties of the samples, and non-destructive testing was used to avoid the dispersion of test data.

(2) During the experiments under different mechanical environments, the energy release process was recorded by the acoustic emission method. And the deformation process was analyzed by the non-contact full-field strain measurement system. Moreover, the particle flow code was used to carry out the quantitative analysis of the microscopic properties. On these bases, the multi-scale evolution process of rock mass under uniaxial compression and hydraulic coupling were compared. Besides, the damage variable was introduced to the evolution process of the dangerous rock mass.

(3) Three sets of experiments were used to analyze the effect of water fluctuations on the rocky reservoir bank. Specifically, the changing trends of macro- and micro-mechanical parameters were obtained through the

weakening experiments under dry–wet cycles and numerical simulation. The influence of the self-weight of upper rock was analyzed by the mechanical state transition under different axial pressures. Moreover, the evolution process considering the time-dependent effect was proposed by the creep experiments under the mechanical state transition.

(4) A new analysis method was proposed to study the evolution process of the dangerous rock masses on the reservoir bank. This method can determine the stability of the submerged dangerous rock mass based on the changing reservoir water. The relevant parameters were also extracted to complete the dynamic collapse analysis after defining the critical state. Furthermore, the mechanical characteristics and the stability trend of the JDRM after the treatment were predicted.

1.4.2 Novelty of the presented research

The innovative points in this book are listed as follows:

(1) Based on the field investigation, indoor test, and in-situ monitoring data, the formation characteristics and the mechanical model of the steep dangerous rock mass on the reservoir bank were analyzed. Furthermore, the buckling failure mechanism of the submerged dangerous rock mass was first proposed.

(2) The influence factors were comprehensively considered, involving the axial pressure, mechanical state transition, and time-dependent effect. And then, the whole evolution process of the rocky bank was first reproduced indoors by the improved multi-scale hydraulic coupling test equipment. Moreover, the evolution model of the dangerous rock mass on the reservoir bank with nonlinear characteristics was obtained.

(3) By introducing the damage accumulated to the transition of mechanical state, a new method was put forward to determine the stability of dangerous rock masses on the reservoir bank. This method can quantify the effects of the reservoir water fluctuation and define the relationship between the damage and the macroscopic strength.

1.4.3 Technical flowchart

In this book, the mechanical model of the dangerous rock mass on the reservoir bank was established according to the field investigation, laboratory tests, and in-situ monitoring data. For the specific geological environment of the high and steep reservoir bank, a comprehensive survey system was proposed. A variety of techniques were used to determine the microscopic mechanical properties of the samples, and non-destructive testing was used to avoid the dispersion of test data. Then, a series of experiments were used to analyze the effect of reservoir water level fluctuations on the rocky bank

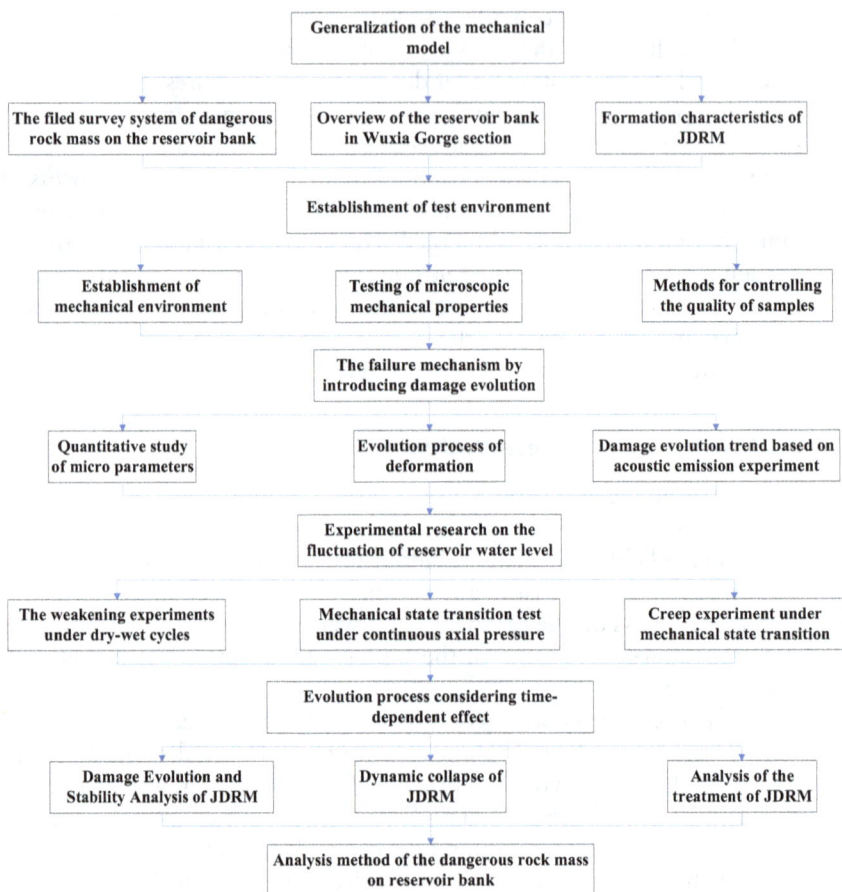

Figure 1.1 Technical flowchart and organization of this book.

through improved existing equipment. According to the test results, a new analysis method was put forward and applied to the JDRM. Furthermore, the stability trend, the collapse simulation, as well as the effect of treatment were analyzed. The technical flowchart, also the content organization of this research is illustrated in Figure 1.1.

Chapter 2

Development of the mechanical model

Wuxia Gorge, located in the east of Wushan County, Chongqing City, is one of the famous Three Gorges on the Yangtze River. The length of Wuxia Gorge in Wushan County is about 24 km. It is a high-steep canyon landform with a height difference from 600 to 1200 m and terrain slope angles from 35° to 75°. The Wuxia Gorge area has experienced three periods of intense geotectonic movement: the Jinning movement in the middle Neoproterozoic, the Yanshan movement in the late Jurassic, and the Himalayan movement in the Neogene. The typical geological structure of Wuxia Gorge is shown in Figure 2.1a. During the recent Himalayan movement, the Wuxia Gorge area experienced intermittent uplifts and tilts that caused multi-level fault cliffs. This violent geotectonic movement further promotes the development of internal joints in the Sinian and Jurassic sediments of the Wuxia Gorge area. Such geological landforms make the cliffs at the trailing edge prone to collapse (Yin et al., 2016; Tang et al., 2019).

According to the geological structure and in-situ investigation of the Wuxia Gorge area, the rocky reservoir bank can be divided into four types: reverse landslide, flat landslide, bedding landslide, and inclined layer landslide (Figure 2.1b; Yin et al., 2023). Under specific structural action, the flat reservoir bank could form the steep dangerous rock mass that is the focus of this book.

Since the storage of the Three Gorges Reservoir area in 2008, with the periodic rise and fall of the water level, a degradation zone of up to 30 m is formed on both banks of the reservoir. The stability of the reservoir bank is reduced due to the changing reservoir water level. On November 23, 2008, the Gongjiafang slope on the north bank of Wuxia Gorge collapsed with a volume of 380,000 m³, which caused a swell of 13 m (Huang et al., 2012, 2014). On June 24, 2015, the Hongyanzi landslide occurred on the left bank of the Daning River in Wushan County, with a scale of about 230,000 m³ (Huang et al., 2016b). The swell caused by this landslide was about 6 m, which affected about 2 km upstream and about 3 km downstream.

DOI: 10.1201/9781003347163-2

(a)

(b)

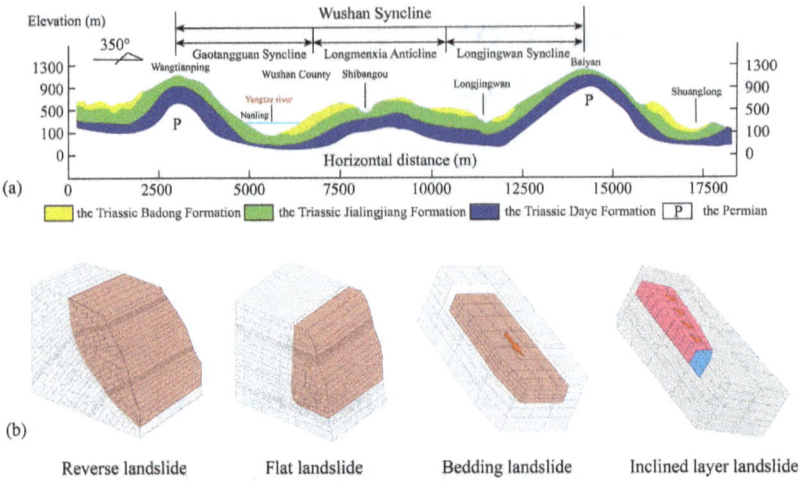

Figure 2.1 The typical geological structure of Wuxia Gorge (a) and the typical rocky bank in the Wuxia Gorge area (b).

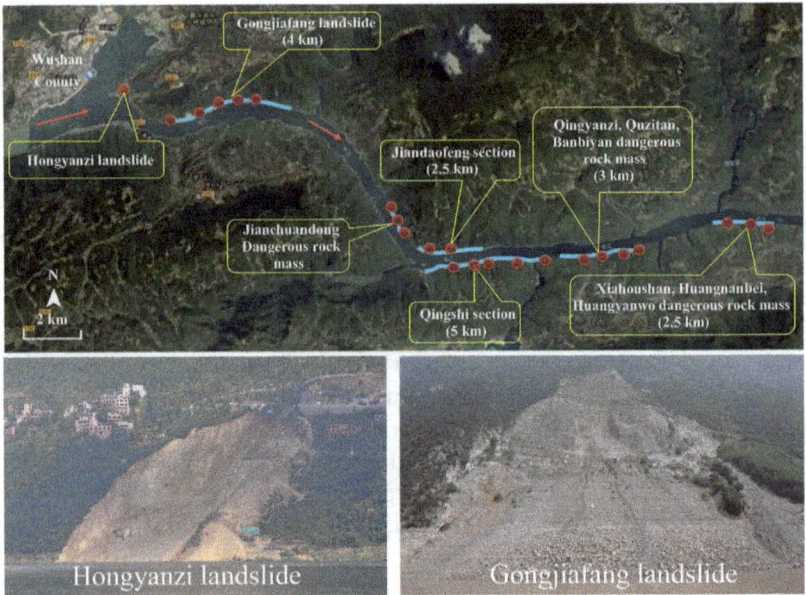

Figure 2.2 The hazardous areas on the reservoir bank.

As of 2020, the reservoir bank has experienced cyclical fluctuations for 12 hydrological years. The hazardous zone induced by the reservoir water fluctuation continues to expand. In the Wuxia Gorge area, the hazardous areas of rocky banks are calibrated (Figure 2.2).

2.2 The survey of dangerous rock mass on the reservoir bank

It is essential to identify the formation characteristics of submerged danger-ous rock masses to control the collapse risk (Feng et al., 2014a, 2014b; Hu et al., 2018; Keneti and Sainsbury, 2018). However, the conventional survey methods that can be applied to areas with high water levels, high mountains, and steep slopes are limited, making it difficult to determine the overall state of submerged dangerous rock masses (He et al., 2019a, 2019b; Kumar et al., 2018; Zhao et al., 2018; Zheng et al., 2018).

Therefore, a comprehensive survey system was proposed to identify the formation characteristics of the dangerous rock mass on the reservoir bank. This system can overcome the limitations of conventional survey methods and provide a comprehensive survey of the steep areas, interior areas, and underwater areas. The technology involved in the investigation system and the corresponding purposes of these techniques are shown in Table 2.1.

In this section, the field survey of Banbiyan dangerous rock mass (BDRM) was taken as a case study to briefly state this survey system and provide a reference for similar dangerous rock masses. The total volume of the BDRM is approximately 735,500 m³, divided into three dangerous rock monomers. The submerged dangerous rock monomer W1 is 718,000 m³ and is the sub-ject of this section. W1 is located in the middle of the bank's slope with a peak elevation of 259.54 m, an underwater basal altitude of 97 m, and a relative height of 162.54 m.

2.2.1 Identifying the boundaries of dangerous rock mass W1

The occurrence of the downstream boundary crack is 318°∠86°, its width is 35–65 cm, its visible depth is 4–6 m, and it extends to the top of the dan-gerous rock mass (Figure 2.3a). The occurrence of the upstream boundary crack is 355°∠63°, its peak elevation is 198 m, and it extends down into the water (Figure 2.3b). The occurrence of the central crack is 32°∠71°, and it extends from the top of the dangerous rock mass to an elevation of 175 m (Figure 2.3c). Due to the geological structure and the unloading of the rock mass, many cracks are developed within the trailing edge of the dangerous rock mass. Based on the 3D oblique photography and the onsite investigations, there are six cracks in the tail of the bank's slope. Among these cracks, the maximum width can reach 56 m (Figure 2.3e). A plurality of transient electromagnetic detection lines was placed on the vertical cliff wall. The geophysical survey results show that the cracks discovered in the field investigations extend into the rock mass. Among them, the third crack (LF3) extends longitudinally through the profile with a depth of 60–90 m, which is defined as the trailing edge boundary crack of the dangerous rock mass (Figure 2.3g).

Table 2.1 The investigation system used to assess submerged dangerous rock masses

Applicable areas	Methods	Limitations of traditional survey methods	Achievement
The steep areas above the waterline	Single-rope technique (SRT)	It is difficult to investigate the cracks in the cliff.	By conducting site surveys of the steep cliffs using SRT, the length, width, and material filling the cracks in the cliff can be determined.
	3D oblique photography	It is difficult to accurately analyze the spatial distribution and boundary characteristics of the dangerous rock mass.	By performing orthographic projection, tilt imaging, and stereo mapping of the dangerous rock mass, its geological conditions and distribution can be determined.
The base of the rock mass underwater	Multi-beam sonar and diving operation	It is difficult to ascertain the underwater terrain and the base of the rock mass.	By scanning the underwater section of the cliff and using diving operations, the shape of the underwater cliff and the development of the rock cavity can be ascertained.
The internal areas of the dangerous rock mass	High-definition camera and wave velocity test in the borehole	It is difficult to assess the corrosion, deterioration, and broken areas within the rock mass.	Using probes and high-definition camera equipment within horizontal boreholes, a continuous image of a cylindrical section of the borehole can be obtained in real time. By using these data along with the results of the wave test, the integrity and dissolution of the rock mass can be determined.
	Geophysical exploration of the cliff	It is difficult to ascertain the development and extension of cracks and broken zones with the bank's slope.	Geophysical exploration methods such as radar and the transient electromagnetic method can be used to collect geophysical data on the submerged cliff, which provides a reference for the development and extension of the cracks and broken areas within the dangerous rock mass.

Source: Yin et al. (2022)

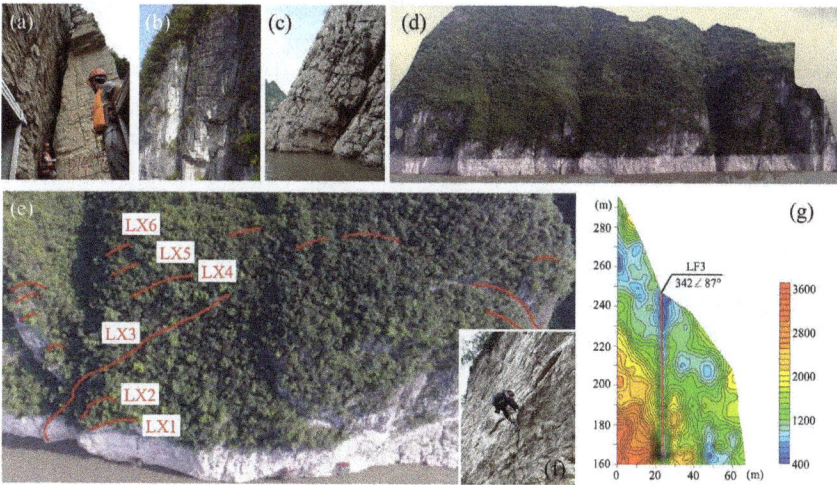

Figure 2.3 Overview of dangerous rock mass W1. (a) Downstream boundary crack. (b) Upstream boundary crack. (c) Central crack. (d) 3D oblique photography. (e) Fracture in the tail of the bank's slope. (f) SRT survey. (g) Typical transient electromagnetic detection profile.

2.2.2   Investigation of the submerged part of the steep cliff

SRT and 3D oblique photography (Figure 2.3d and 2.3f) were used to investigate the cracks and broken areas in the submerged part of the steep cliff. A total of 33 cracks and six broken areas are identified in the survey (Figure 2.4a). According to statistical analysis, the cracks occur at 52–118°∠67–88°, and the cracks are 10.22–93.86 m in length. Moreover, due to the fluctuation of reservoir water, the development degree of the cracks in the degradation zone is higher than that of the cracks in the upper zone.

2.2.3   Formation characteristics of the base of the rock mass

The base rock mass is key to the overall deformation and failure processes of the dangerous rock mass. A total of four deep horizontal drilling holes were placed in the degradation zone (Figure 2.4a). Meanwhile, the internal geological conditions of the boreholes were examined using high-definition camera technology. Besides, the wave velocity tests were conducted in the drilling hole. Using borehole XZK2 as an example, the results can be obtained as follows. There are significant cracks at 7.5 m and 23.4–23.6 m (Figure 2.4b). From 39.3 m to 42.8 m, the surface is rough, and the mud

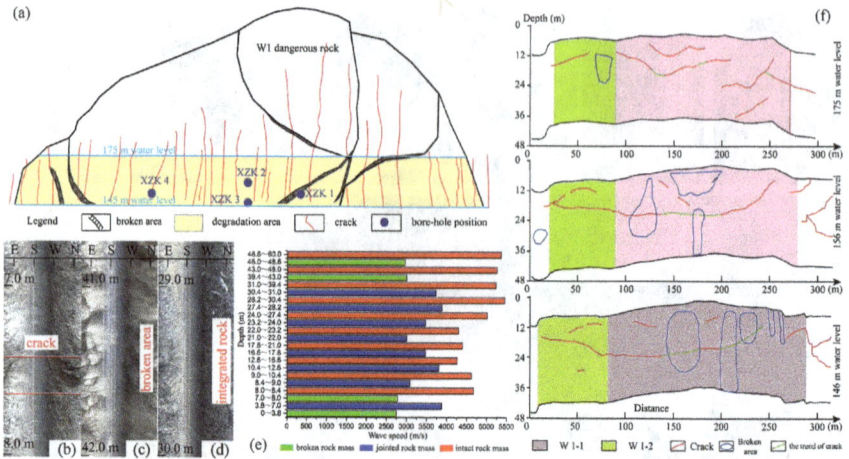

Figure 2.4 Investigation of the submerged part of the steep cliff. (a) Cracks and broken areas. (b–e) Drilling results in the degradation zone. (f) Interpretation profile from geological radar.

composition is more massive; thus, this part is determined as the broken area (Figure 2.4c). The surface of the remaining section is smooth, which indicates that the rock mass is integrated (Figure 2.4d). The results of the wave speed tests are consistent with the results of the high-definition camera technology (Figure 2.4e).

Three geological radar test lines were completed at elevations of 146 m, 156 m, and 175 m on the cliff wall, and the internal geological conditions of the corresponding elevations were interpreted (Figure 2.4f). Combined with the survey mapping and drilling results, it can be found that the continuous length of the upstream crack is about 70 m, and the rock mass separated by the cracks gradually thickens from upstream to downstream. At lower elevations, the broken areas' development degree is higher, and they tend to connect to each other. At higher elevations, the broken areas' development degree is lower, while the cracks' development degree is higher.

Since the base rock mass extends underwater, it is very difficult to carry out investigations. There are six obvious broken areas within the cliff wall above the 145 m water level. Based on the results of the underwater multibeam sonar and diving surveys (Figure 2.5a and b), two broken areas connect each other underwater. And the development elevations of broken areas 2 and 3 are 134.50 m and 97.57 m, respectively (Figure 2.5c).

2.2.4 Results

Based on the previous investigation, it can be found that the trailing edge of the boundary crack in W1 is LF3, and W1 can be divided into two subareas

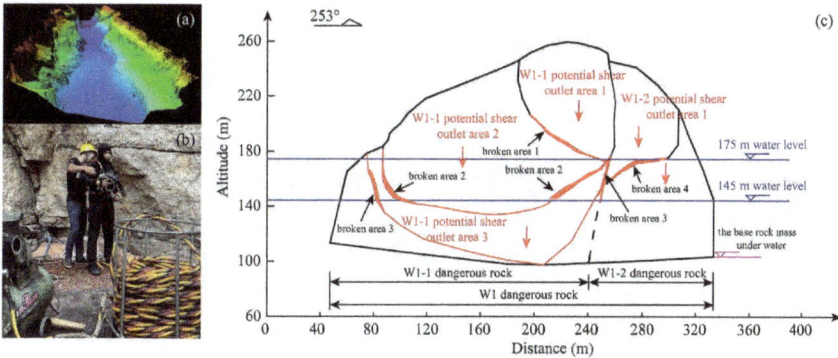

Figure 2.5 (a) Scanning results of the multi-beam sonar. (b) Diving operation. (c) Comprehensive analysis of W1.

(W1–1 and W1–2) by the central crack (Figure 2.5c). Due to the development of broken areas 1–3, subarea W1–1 has three potential shear outlet areas. Whereas, broken area 4 is the potential shear outlet area of subarea W1–2. Specifically, subarea W1–1 has a width of 193.0 m, an upper elevation of 259.54 m, a lower basal elevation of 97.57 m, a thickness of 15–54 m, and a potential shear-out volume of about 322,700 m³. The subarea W1–2 has an upper elevation of 247.32 m, a minimum base elevation of 134.5 m, a thickness of 6.5–35.47 m, and a potential shear-out volume of 259,000 m³.

2.3   Formation characteristics and the mechanical model of the JDRM

As the monitoring data of the JDRM was sufficient, the JDRM was selected as the research focus of this book. The laboratory test and monitoring data were used to analyze the formation characteristics of the JDRM.

2.3.1   Overview of the JDRM

The JDRM is located on the left bank of the Yangtze River and the Wuxia Gorge of the Three Gorges Reservoir Area (E 110°00′01″, N 31°01′35″). The upstream boundary of the JDRM is a longitudinal crack (Crack 1) that runs from the gully (226 m) to the foot of the slope (150 m). The boundary crack on the downstream side (Crack 2) is clearly visible on the steep cliff surface. The crack is wider at the top and narrower at the bottom, is filled or partially filled with gravel soil or dissolved residual gravel soil, and gradually migrates downward to an elevation of 153 m. The strike of the upper part of Crack 2 is 324°∠65°, the middle part is turned upright and gently inclined, and the strike of the lower part of the crack has an orientation of

Figure 2.6 **The Goddess Peak anticline and the front view of the JDRM.**

324°∠45°. The adjacent steep slope represents the JDRM base rock mass. The JDRM has four unloading cracks (Cracks 3–6), which extend from the steep cliff to the trailing edge and have strikes of 276°–260°∠75–85°. Among these unloading cracks, Crack 3 (the trailing-edge boundary of JDRM) is the deepest crack (depth of 3.15 m); it is filled with gravel below an elevation of 226 m. The other cracks (Cracks 4–6) are only open at the top of the rock mass, and their opening depth is shallow. The front view of the JDRM is shown in Figure 2.6.

Boundary cracks have given the JDRM the shape of an irregular hexahedron. The elevation of the trailing-edge boundary ranges from 278 to 305 m, the elevation of the foot of the rock mass is 155 m, and the average height difference between these two boundaries is 135 m. The average width of the dangerous rock mass is about 55 m, and the average thickness is about 50 m. The volume of the JDRM is approximately 36×10^4 m^3, and its main collapse direction has an orientation of 260°. The JDRM is primarily composed of the fourth section of the Triassic Daye Formation (T_1d^4), the first section of the Triassic Jialingjiang Formation (T_1j^1) (only above 280 m), and the third section of the Triassic Daye Formation (T_1d^3) (the base of the rock mass). A typical cross-section of the JDRM is shown in Figure 2.7a, which includes the location of a pressure sensor that resides at the bottom of an adit. During the worst Yangtze River flooding in 50 years, the water level in the reservoir area reached 162.1 m (Yin et al., 2016; Tang et al., 2019); the analyses in this book focused on the highest (175 m) and lowest (145 m) water levels in the reservoir area, as well as the periodic fluctuation of the water level over time.

2.3.2   Formation characteristics of the JDRM

The upstream and downstream boundaries of the JDRM are the same set of structural planes, in line with the axis of the Goddess Peak anticline. Meanwhile, several large-scale structural planes in the dangerous rock area intersect the Goddess Peak anticline in a near radial pattern, and these directions are also nearly consistent with the axis of the anticline (Figure 2.6).

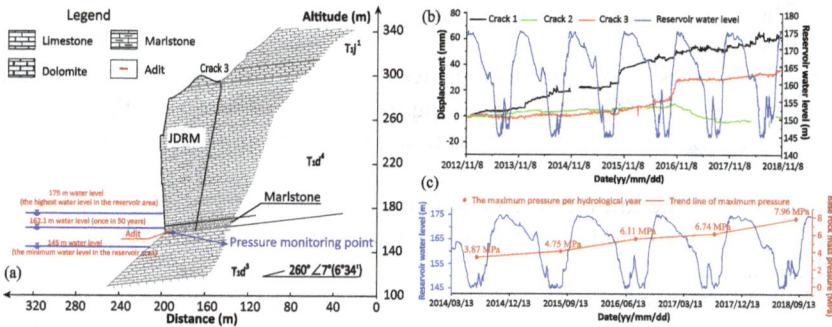

Figure 2.7 (a) Cross-section of the JDRM. T_1d^3 is the third section of the Triassic Daye Formation, T_1d^4 is the fourth section of the Triassic Daye Formation, and T_1j^1 is the first section of the Triassic Jialingjiang Formation. (b) Monitoring data of boundary cracks. (c) The pressure experienced by the base rock mass.

The JDRM was formed by the unloading of the bank slope, erosion, and downcutting driven by the Yangtze River, preexisting joints and fractures, and the release of structural stress. After its formation, the base rock mass of the JDRM continues to be deformed by the weight of the upper rock mass and the fluctuation of reservoir water. Cracks developed rapidly as the base rock mass weakened and accumulated more damage. According to the monitoring data (Figure 2.7b and c), the JDRM is under significant uneven pressure resulting in a yearly increase in the width of boundary cracks, stress, and deformation in the direction of the lower reaches of the Yangtze River. The fluctuation of the reservoir water level accelerates the deterioration of the base rock mass and the overall instability of the JDRM.

In addition to conducting field investigations and on-site monitoring, the marlstone samples were collected from the JDRM base rock mass for dry–wet cycle testing in 2014. After accounting for existing experimental specifications (Specification for rock tests in water conservancy and hydro-electric engineering, 2001; Standard for test methods of engineering rock mass, 2013; Ulusay, 2014), the site conditions (the maximum temperature of the rock mass in the reservoir area could reach 60°C), and the relevant published works (Hudson and Harrison, 2000; Hua et al., 2017; Zhou et al., 2017b; Liu et al., 2018a), the three-step dry–wet cycle test methodology can be devised:

(1) The free water immersion method was used to achieve the process of saturation. After the samples were oriented vertically in a container, water was added initially to a height of h/4 (h is the height of the sample). Every two hours for the next six hours, the water height was increased by h/4 until the samples were completely submerged.

(2) The samples were removed from the container after they had been sub-merged for 48 hours.
(3) The samples were placed in an oven for 48 hours with the drying tem-perature set to 60°C. After drying, it was determined that the moisture content of the samples was less than 0.1% (i.e., approximately zero).

After running the samples through 5, 15, 20, and 30 dry–wet cycle tests, the experiments were conducted to obtain the sample mechanical parameters (Table 2.2).

Based on the experimental results, for the samples in their natural state that experienced thirty dry–wet cycles, the uniaxial compressive strength, and the elasticity modulus had decreased by 21.45% and 41.23%, respectively. For the samples in their saturated state that experienced thirty dry–wet cycles, the uniaxial compressive strength and the elasticity modulus had decreased by 26.96% and 42.19%, respectively. Meanwhile, the tensile strength, friction angle, and cohesion decreased by 28.18%, 12.88%, and 17.85%, respec-tively. And the Poisson's ratio increased by 16.67%.

2.3.3   Generalization of the mechanical model

Based on the formation characteristics of JDRM, the generalized JDRM model can be simplified (Figure 2.8). The periodic change in the water level has significantly damaged the marlstone zone at the foot of the JDRM. Com-pared to the marlstone at the base of the JDRM, the middle-upper part of the JDRM is robust and has fewer cracks. The mechanical analysis shows that the weight of the middle-upper rock mass is roughly equivalent to the axial compressive pressure; as such, the marlstone zone can be summarized with a damage model that incorporates long-term axial stress and the fluctuation of reservoir water level.

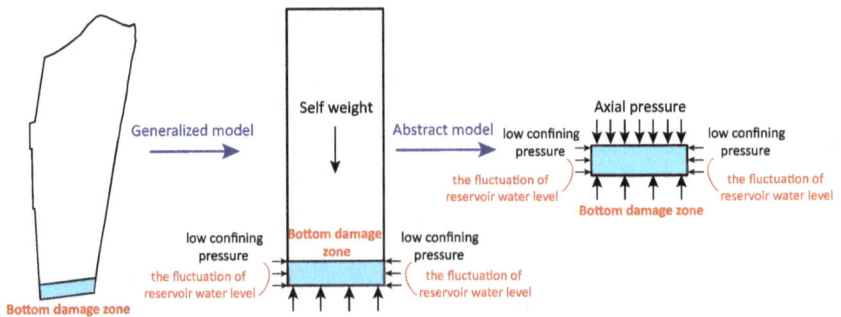

Figure 2.8 Generalization of the JDRM mechanical model.
Source: Wang et al. (2020c)

Table 2.2 Mechanical parameters of marlstone from the Three Gorges Reservoir Area after experiencing different numbers of dry–wet cycles

Number of dry–wet cycles	Uniaxial compressive strength (Natural state, MPa)	Uniaxial compressive strength (Saturated state, MPa)	Tensile strength (MPa)	Friction angle (°)	Cohesion (MPa)	Elasticity modulus (Natural state, ×10⁴ MPa)	Elasticity Modulus (Saturated state, ×10⁴ MPa)	Poisson's ratio
0	19.07	13.24	1.10	32.6	3.36	0.405	0.391	0.30
5	18.39	12.25	0.99	32.2	3.16	0.373	0.358	0.31
15	16.96	11.10	0.89	31.8	2.91	0.351	0.327	0.32
20	16.22	10.44	0.83	30.1	2.83	0.272	0.261	0.33
30	14.98	9.67	0.79	28.4	2.76	0.238	0.226	0.35

Source: Wang et al. (2020d)

Specifically, three problems need to be studied in the analysis of damage evolution:

(1) Transformation of time scale. The strength reduction involved in the general dry–wet cycle test is quantified yearly. However, the accumulation of damage in the reservoir bank continues over time.
(2) Construction of hydraulic coupling test conditions. In the case of constant or increasing axial pressure, the rocky bank gradually transfers from dry state to saturated state and then to certain water pressure.
(3) The continuity of mechanical state transition. In each mechanical state, there is sufficient time to complete the stress adjustment that can form multiple-stage stress balance.

The related issues will be studied through improved experimental equipment, and the detail will be listed in Chapter 3.

2.4   Summary of this chapter

The geological structure of the Wuxia Gorge area is complex and the cliffs at the trailing edge are prone to collapse. Since water storage in the reservoir area in 2008, the fluctuation of reservoir water level has become an essential factor inducing the instability of the bank slope. And the hazardous zone caused by the reservoir water fluctuation continues to expand over time.

A comprehensive survey system was constructed using various technologies and methods, which can effectively detect the actual state of a submerged dangerous rock mass in a steep valley. Respectively, the development of cracks and broken areas in cliffs, underwater sections, and the interior of the dangerous rock mass can be determined by this survey system. The related survey results can provide active support for the judgment and analysis of the formation and failure mechanisms of dangerous rock masses.

Based on the filed survey, monitoring data, and indoor test, the mechanical model of JDRM was generalized. It indicates that the self-weight of the upper rock mass and the deterioration of the base rock mass caused by the reservoir water fluctuation are the main factors leading to the deformation and failure of the JDRM. This mechanical model is the basis of experimental research.

Chapter 3

Establishment of the test environment

3.1 Establishment of the mechanical environment

A multi-scale mechanical experimental system (HMC-1000–60), developed by Sun Yat-sen University, was used to study the damaging effect of rock bearing the hydromechanical coupling action (Figure 3.1). This system can simulate the whole process of hydro-mechanical coupling damage of rock under the conditions of high dynamic or static water pressure and other working fluids. Besides, it can carry out a variety of rheological damage tests with different solutions, pressures, and rheological stages. In the same period, both high-precision non-destructive measurement and multi-scale observation can be achieved. A detailed description of this apparatus can be referred to in Liu et al. (2018a) and Zhou et al. (2019).

Figure 3.1 Multi-scale mechanical experimental system for rheological damage effect of rock bearing the hydro-mechanical coupling action.

Source: Wang et al. (2020c)

DOI: 10.1201/9781003347163-3

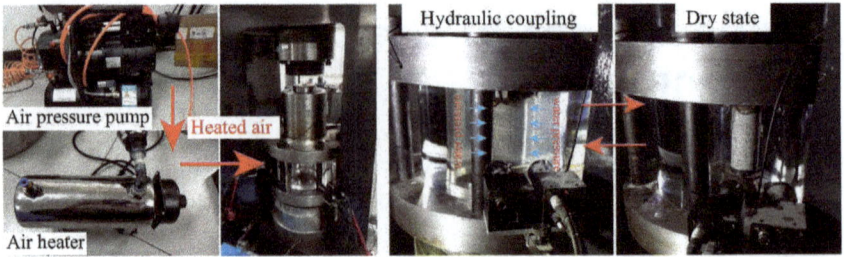

Figure 3.2 The transition of mechanical state under continuous axial pressure.

For the subject studied in this book, the most significant advantage of this equipment lies in the construction of hydraulic coupling. Specifically, water is directly used to apply confining pressure to the rock mass, which can highly restore the water-stress environment of the base rock mass on the reservoir bank.

To further realize the mechanical environment of the rocky bank, a drying module was added to the original equipment. This module can ensure the sample completes the conversion in the dry state and the hydraulic coupling state while the axial pressure continues to act. The mechanical environment constructed by the experiment system is shown in Figure 3.2.

According to the mechanical model proposed earlier, the following experimental research was designed through this improved test system:

(1) The effect of hydraulic coupling and the entire evolution process of rock were explored through compression experiments in different mechanical states.
(2) The deterioration trend of the rock mass under mechanical state transformation was studied through the dry–wet cycle test.
(3) The influence of the self-weight of upper rock on reservoir bank was investigated through the mechanical state transition test under the continuous action of axial pressure.
(4) The time-dependent effect on rock mass was analyzed through the creep test under the transition of the mechanical environment.

Since the treatment of the JDRM already began during these experiments, it is very difficult to obtain the marlstone in the base area. The samples taken from the foot of the JDRM are the limestone of the Jialingjiang Formation. The polarizing microscope, Scanning Electron Microscope (SEM) method, and Energy Dispersive Spectrum (EDS) method were used to evaluate the microscopic mechanical properties of the selected samples qualitatively. And the quality of the sample was controlled to reduce the dispersion of the test data using computed tomography (CT) scanning and wave speed testing.

3.2   Testing of mechanical properties

3.2.1   Analysis of polarizing microscope

The observation of the polarizing microscope was completed at the School of Earth Resources of China University of Geosciences (Wuhan). The test instrument is the Olympus-BX51 polarizing microscope produced by Japan (Figure 3.3a). The rock is mainly composed of transparent minerals. And the polarizing microscope is an indispensable instrument for studying birefringent substances by using the polarization characteristics of light. The crystal form, cleavage, and other features of the mineral can be observed under a polarizing microscope (Chang et al., 2006).

The thin slices of rock were obtained through cutting, smoothing, gluing, slicing, grinding, and polishing. After processing, the thickness of the slices was around 30 μm to ensure the light source pass through (Figure 3.3b). As the light features varied with different types of minerals, the mineral's optical properties can be used as an essential basis for mineral identification.

Observation by polarizing microscope (Figure 3.4) showed the microstructure of limestone under the single polarized light and orthogonal polarized light.

By analyzing the sample slices under plane-polarized light and perpendicular polarized light, it can be found that the selected sample's mineral

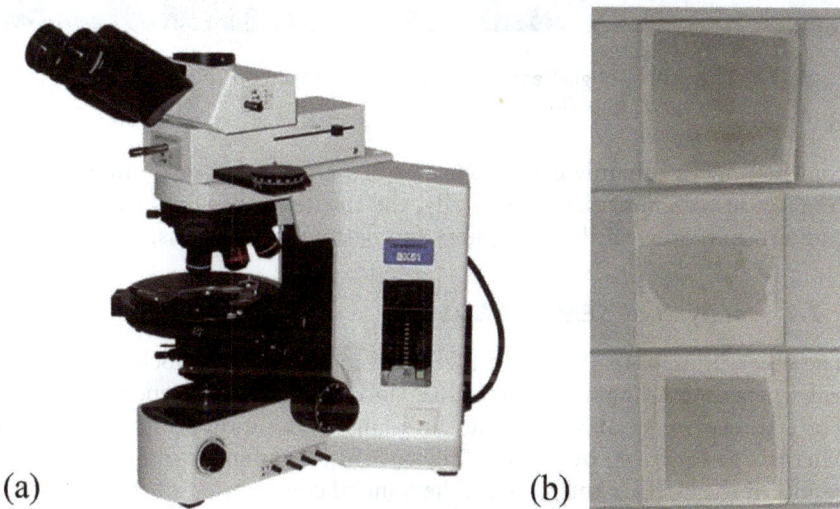

(a)   (b)

Figure 3.3 Analysis of polarizing microscope. (a) Olympus-BX51 polarizing microscope. (b) The thin slices of rock.

Figure 3.4 Observation results of the polarizing microscope (magnify 20 times). (a) Plane polarized light. (b) Perpendicular polarized light.

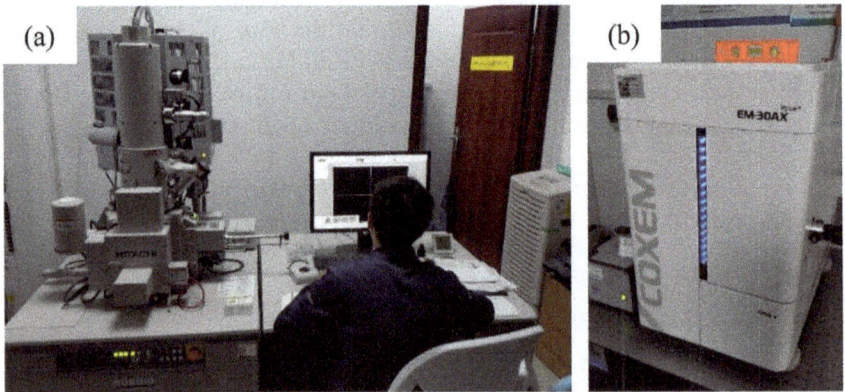

Figure 3.5 The Ultra-High-Resolution Scanning Electron Microscope (a) and EM-30AX plus (b).

composition was mainly calcite which was a typical carbonate mineral and contains fewer impurities. Specifically, the calcite in the slices was colorless and transparent, with rhombic joints and flashing protrusions.

3.2.2 Analysis of SEM and EDS

The observation of the SEM and EDS were completed at the School of Material Science and Chemistry of China University of Geosciences (Wuhan). The test instrument of SEM is the Ultra-High-Resolution Scanning Electron Microscope produced by Hitachi (Figure 3.5a). Combined with the scanning results of the electron microscope, the mineral composition of the rock mass can be analyzed by the test equipment of EDS (EM-30AX plus; Figure 3.5b). Specifically, the SEM method can generate an image of the sample surface by scanning a focused electron beam. And the interaction between the electrons and the atoms in the sample can produce various signals containing surface morphology (McMullan, 2006). The EDS method is mainly for qualitative and quantitative analysis of the chemical composition of the material micro-area (Braga et al., 2002). Its characterization capabilities are due in large part to the fundamental principle that each element has a unique atomic structure allowing the unique set of peaks on its X-ray spectrum. Before conducting the test, it was required to spray gold on the surface of the sample to improve its conductivity (Figure 3.6).

The SEM results (Figure 3.6) indicated that the surface structure of the sample was dense without obvious microscopic pores and impurities. This phenomenon was consistent with the results of the polarizing microscope.

For an area on the sample (Figure 3.7), EDS was performed to obtain the distribution map of different elements. It can be found that the sample was mainly composed of three elements: carbon, oxygen, and calcium.

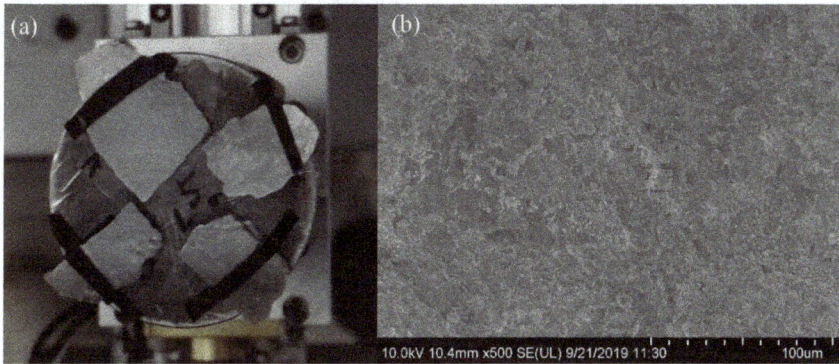

Figure 3.6 The sample after gold spraying (a) and the result of SEM (b).

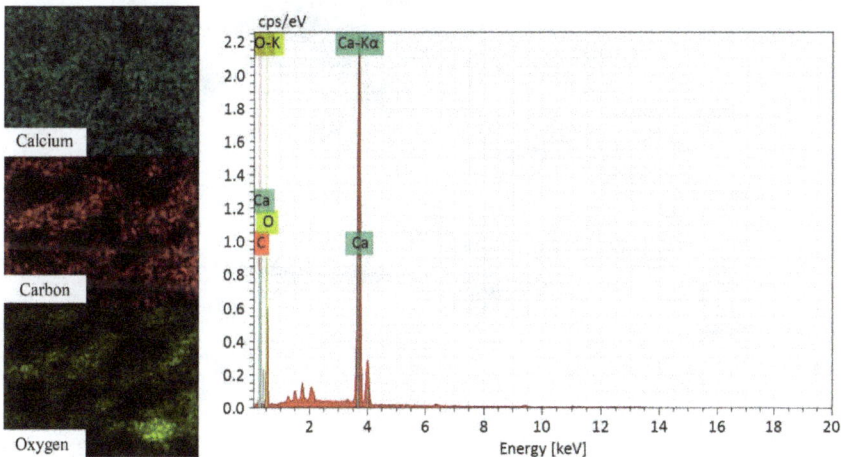

Figure 3.7 The energy distribution of different elements.

According to the quantitative analysis of the content of different elements, the content of the carbon element was 8.2%, the content of the oxygen element was 50.25%, and the content of the calcium element was 36.7%. The total content of these three elements was 95.07%. Therefore, it can be determined that the constituent minerals of this sample were mainly calcite and a few other components.

3.3   Methods for controlling the quality of samples

The CT scan is an imaging procedure that uses computer-processed combinations of many X-ray measurements taken from different angles to produce cross-sectional (tomographic) images of specific areas of a scanned object, allowing the user to see inside the object without cutting (Herman, 2009). Therefore, CT scanning can realize non-destructive testing and three-dimensional visualization to obtain the rock sample's internal structure fully. The equipment used here is Perspective Non-destructive High-resolution Nano Voxel-4000 that is produced by Tianjin Sanying Precision Instrument Co., Ltd (Figure 3.8).

According to different gray levels, the substances involved in the CT data can be distinguished. A slice chart in XY direction was selected randomly

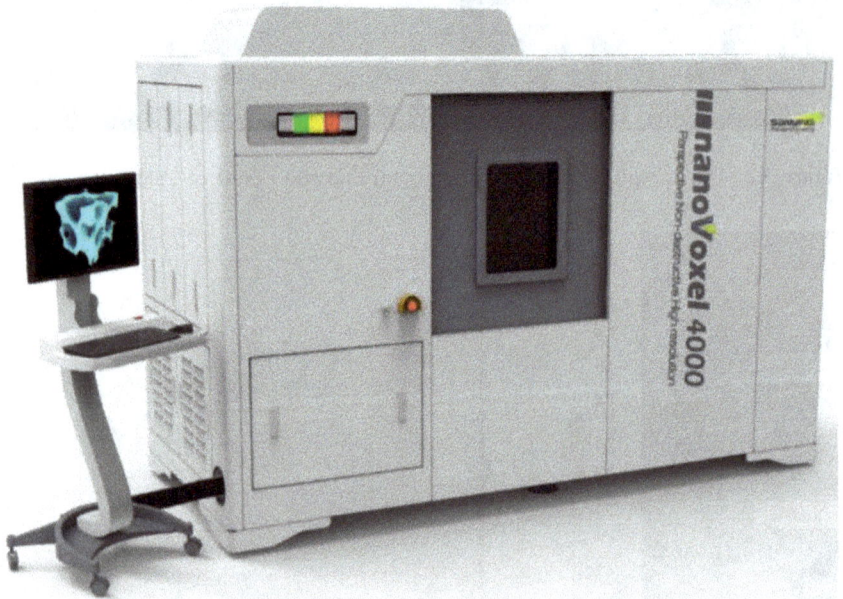

Figure 3.8 The Nano Voxel-4000 equipment.

and a straight line on the picture was highlighted to get the corresponding density change curve. It can be found that point A was a low-density pore and appeared like a valley of waves; point B was a high-density material and appeared as a peak of waves.

The two-dimensional slice can better identify the distribution of mineral particles and pores in different directions (Figure 3.9). Furthermore, the cracks, pores, and minerals in the sample can be measured. The results showed that the pores varied in size, and these pores' distribution was very uneven.

Using the threshold to extract the pores (Figure 3.10a), the porosity of the sample was calculated to be 0.008%. The layer-by-layer porosity in the Z direction indicated that the cut surface's porosity was maintained between 0% and 0.02%, with an average porosity of 0.008% (Figure 3.10b). The minerals were extracted by threshold segmentation, and their volume percentage was determined to be 0.017% (Figure 3.10c).

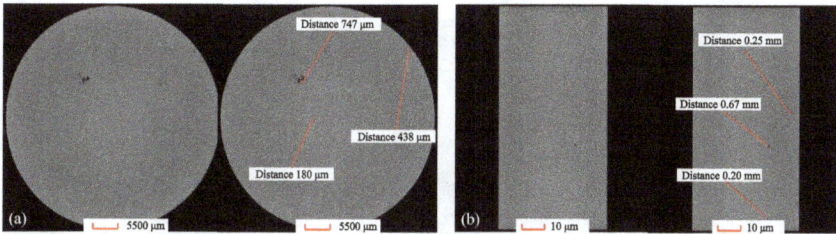

Figure 3.9 Determination of pores in the different directions. (a) Determination of pores in the XY direction. (b) Determination of pores in the YZ direction.

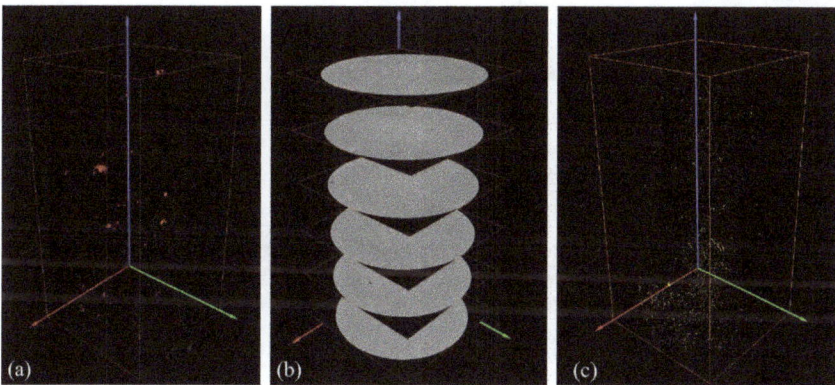

Figure 3.10 The extraction results of the sample. (a) The distribution of pores. (b) The extraction of pores in the Z-direction. (c) The distribution of minerals.

Through further statistics of the pores (Figure 3.11a), it can be found that the equivalent pore diameter was less than or equal to 400 μm.

Combined with the results of CT scanning, the distribution characteristics of pores and minerals in three-dimensional space can be better understood (Figure 3.11b). There were few pores and impurities in the selected sample, and these pores were difficult to form an active seepage channel.

Besides, the test of wave velocity was used to ensure that the sample's internal structure was similar (Figure 3.12). The equipment used here is a holographic acoustic emission signal analyzer produced by Beijing Softland Times Scientific and Technology Co. Ltd. The wave velocity of the sample was about 5700 m/s. Before the test, the samples were selected to ensure that the wave speed error was below 5%. Meanwhile, the wave velocity will be used as the reference parameter of the acoustic emission experiment (Xue and Zhang, 2011).

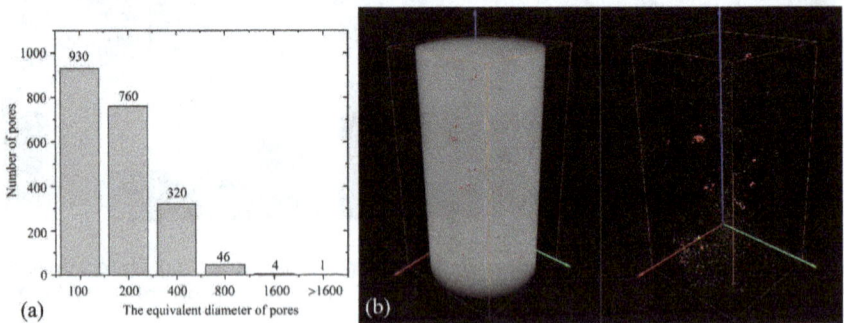

Figure 3.11 (a) The statistical analysis of pores. (b) Comprehensive results of CT scan.

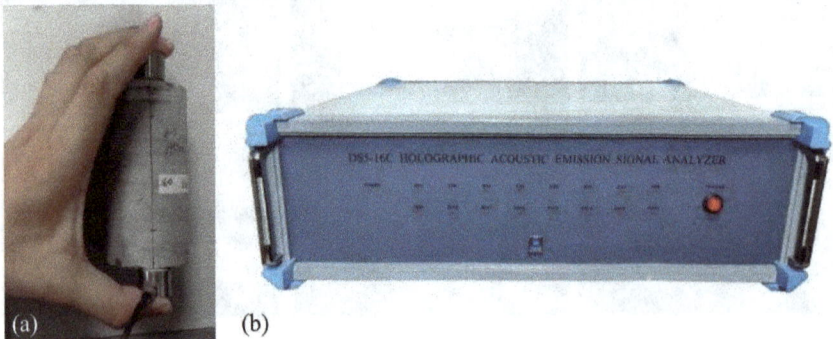

Figure 3.12 The test of wave velocity (a) and the holographic acoustic emission signal analyzer (b).

3.4   Summary of this chapter

According to the mechanical model of dangerous rock mass on the reservoir bank, an improved multi-scale mechanical experimental system was used to establish the complex mechanical environment. Through this improved experimental system, the samples can complete the conversion in the dry state and the hydraulic coupling state while the axial stress continues to act. Under the test equipment, four sets of tests were set up to study the hydraulic coupling, dry–wet cycles, axial pressure, and time-dependent effect. The relevant experimental content will be described in detail in Chapters 4 and 5.

The polarizing microscope, SEM and EDS were used to evaluate the microscopic mechanical properties of the selected samples qualitatively. The results indicated that the constituent minerals of this sample were mainly calcite with few other impurities.

Through CT scanning and wave speed testing, the quality of the sample was controlled to reduce the dispersion of the test data. The results of the CT scanning showed that the selected samples had few pores and impurities, and these pores were difficult to form the active seepage channel. The wave velocity of the sample was about 5700 m/s which would be used as the reference parameter of the acoustic emission experiment.

Chapter 4

Analysis of hydraulic coupling test

In this chapter, the experimental analysis under different mechanical environments was conducted to study the influence of hydraulic coupling on rock mass. Based on the test results, the quantitative analysis of the microscopic properties was also carried out by the particle flow code (PFC). Combined with the case study of the Zengziyan collapse, the entire evolution of the dangerous rock mass under different mechanical state were compared and analyzed. Moreover, the collapse characteristics of the submerged dangerous rock mass were clarified to provide a basis for further analysis in subsequent chapters.

4.1 Analysis of macro-strength

Based on the mechanical model proposed in Chapter 2 and the mechanical environment established in Chapter 3, the rock masses under five kinds of mechanical states were tested and analyzed. These five mechanical environments included uniaxial compression under dry state, uniaxial compression under saturated state, hydraulic coupling under confining pressure 0.3 MPa, hydraulic coupling under confining pressure 1 MPa, and hydraulic coupling under confining pressure 3 MPa.

Using the test control method in Chapter 3, three samples were tested in each mechanical state, and the typical test curves under different test conditions were compared for analysis. Since the water level in the Three Gorges Reservoir area fluctuates between 145 m and 175 m, the maximum water level difference is 30 m which corresponds to a water pressure of 0.3 MPa. Therefore, this study took the water pressure of 0.3 MPa as a focus. The confining pressure was then increased by three times, and the hydraulic coupling states under 1 MPa and 3 MPa were further analyzed. The comparison of different test conditions is shown in Figure 4.1.

The stress–strain curves under different mechanical states are shown in Figure 4.2. According to the test results, the following conclusions can be obtained.

(1) For the dense limestone used in the test, the saturated state's uniaxial compressive strength was 5.10% lower than that in the dry state, and

DOI: 10.1201/9781003347163-4

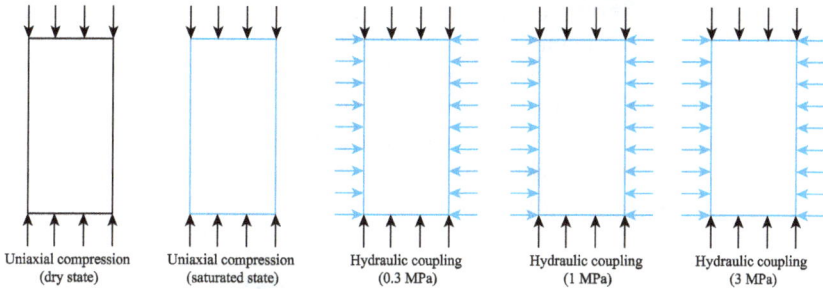

Figure 4.1 Comparison of test conditions.
Source: Zou et al. (2022)

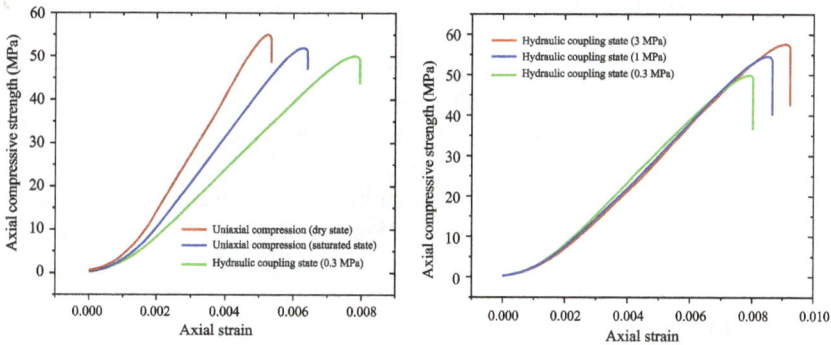

Figure 4.2 The stress–strain curves under different mechanical states.
Source: Zou et al. (2022)

this decrease was not significant. It indicated that the saturated state had less effect on the strength of dense limestone. However, the strain corresponding to the saturated state's peak strength was higher than that of the dry state, indicating that the plasticity of the limestone increased in the saturated state. This difference in plastic characteristics may be amplified when the rock mass is converted between dry and saturated states. Related experiments will be further discussed in Chapter 5.

(2) Under the hydraulic coupling mechanical conditions of low confining pressure (0.3 MPa), the strength of limestone will be further reduced compared with the saturated state. Specifically, the strength of limestone in the 0.3 MPa hydraulic coupling state was 3.64% lower than that in the saturated state, and 8.56% lower than that in the dry state. This phenomenon revealed that under hydraulic coupling, the participation of water in the evolution process increased. And the hydraulic coupling not only softened the rock's particles but also accelerated the failure process through the existing microdamage.

(3) When the confining pressure continued to increase, the peak strength of limestone gradually increased. In terms of actual working conditions, the peak strength of the rock mass in the deep-water area was higher than that in the shallow-water area. Although the peak strength of the rock mass increased, the amount of plastic deformation continued to grow. The relevant analysis will be further studied after obtaining the microscopic parameters by numerical simulation.

4.2 Quantitative study of micro-parameters

The PFC was used to analyze the micro-parameters under different mechanical states quantitatively, which can simulate the movement and interaction of finite-sized particles. These particles are defined as rigid elements with finite mass that move independently of one another and can both translate and rotate. Particles interact at pair-wise contacts through an internal force and moment. Contact mechanics is embodied in particle interaction laws that update the internal forces and moments. The time evolution of this system is computed via the distinct-element method, which provides an explicit dynamic solution to Newton's laws of motion. The PFC model provides a synthetic material consisting of an assemble of rigid grains that interact at contacts and include granular and bonded materials (Cundall, 1971; Hart et al., 1988; Potyondy, 2015).

Compared with several common numerical simulation methods (Finite Element method, Distinct Block Element code, and Fast Lagrangian analysis), the PFC can break through the limitation of deformation, and conveniently clarify the mechanical problems of discontinuous material. Besides, it can reflect the different physical relationships of multiphase media, and effectively simulate the continuous phenomena such as cracking and separation of media.

In this book, the parallel bond model, mostly applied to simulate the materials such as rock and concrete, was used to define the contact relationship between particles in the numerical calculation. Besides, this bond model has the characteristics of high bond strength and can withstand bending moments and other loads.

4.2.1 Introduction of linear parallel bond model

The linear parallel bond model provides the behavior of two interfaces: an infinitesimal, linear elastic (no-tension), and frictional interface that carries a force and a finite-size, linear elastic, and bonded interface that carries a force and moment (Figure 4.3). The first interface is equivalent to the linear model: it does not resist relative rotation, and slip is accommodated by imposing a Coulomb limit on the shear force. The second interface is called a parallel bond, because when bonded, it acts in parallel with the first interface. When the second interface is bonded, it resists relative rotation, and

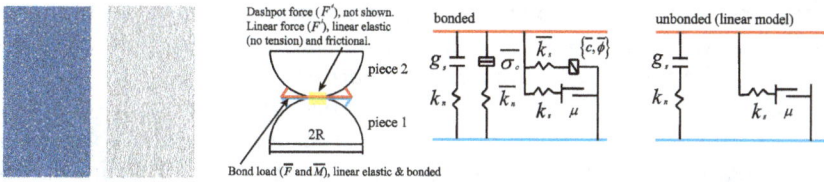

Figure 4.3 Behavior and rheological components of the linear parallel bond model with inactive dashpots.

Source: Zou et al. (2022)

its behavior is linear elastic until the strength limit is exceeded and the bond breaks, making it unbonded. When the second interface is unbonded, it carries no load. The unbonded linear parallel bond model is equivalent to the linear model.

Contact with the linear parallel bond model is active if it is bonded or if the surface gap is less than or equal to zero. The force-displacement law is skipped for inactive contacts. When the reference gap is zero, the notional surfaces of the first interface coincide with the piece surfaces.

4.2.2 Calibration and sensitivity analysis of micro-parameters

According to the macro test data, the numerical simulation parameters were calibrated through multiple corrections. And then the numerical model can realistically reflect the loading process of the sample and link the macro-mechanical parameters to the micro-parameters. Besides, based on the constitutive relationship of the parallel bond model, the calibration of micro-parameters followed these principles:

(i) The parallel bond model degenerated into a linear contact model after the bond was broken. Therefore, when the specimen was under tension or compression, the parallel bond part worked, while its deformation modulus was different. And the linear contact part only worked when the sample was pressed.

(ii) The Poisson's ratio of elastic deformation was affected by the normal-to-shear stiffness ratio, and these two parameters were linearly related.

(iii) The value of effective bond modulus was determined by the elastic modulus, and these two parameters were linearly related.

(iv) The failure mode of the specimen was determined by the normal-to-shear stiffness ratio.

(v) After the bond normal-to-shear stiffness ratio was determined, the amplification factor of the microscopic bonding parameter was linearly related to axial compressive strength.

In this part, the uniaxial compression experiment was used as an example to analyze the calibration process of micro-parameters in detail, which provided a basis for the subsequent parameter comparison.

When the uniaxial compression test was performed in the dry state, the elastic modulus of the sample was 12.5 GPa; the tensile modulus was set to 9.615 GPa (Perkins and Krech, 1968; Hawkes et al., 1973; Haimson and Tharp, 1974; Stimpson and Chen, 1993); the Poisson's ratio was 0.25; and the uniaxial compressive strength was 55 MPa. The micro-parameters will be calibrated based on these macro-parameters.

4.2.2.1   Establishment of the numerical model

In the process of constructing the numerical model, the porosity was set to 0.01; the diameter of the particle unit was set from 0.6 mm to 0.8 mm; and then a two-dimensional model with a diameter of 50 mm and a height of 100 mm was obtained. There were 3186 particles and 8078 initial contacts in this numerical model. The same physical model was used to analyze the micro-parameters in different mechanical states quantitatively.

4.2.2.2   Calibration of the bond effective modulus

The effective modulus was set to a small value (0.1 MPa), and the bond effective modulus (Pb_emod) was changed to 1 GPa, 5 GPa, 10 GPa, and 20 GPa. Meanwhile, the other parameters were set to relatively high values. Based on the corresponding stress–strain curve by numerical tensile test, the relationship between the tensile modulus (the ratio of peak intensity to peak strain) and the bond effective modulus can be obtained through the fitting (Figure 4.4).

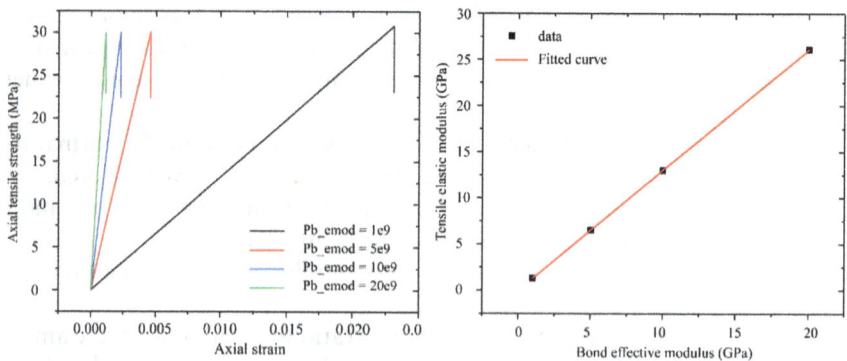

Figure 4.4 Tensile test curves (left) and parameter fitting (right) under different bond effective modulus.

According to the experimental results of numerical simulation, the linear relationship between the bond effective modulus (Pb_emod) and the tensile modulus (E_t) is:

$$E_t = 1.3031 \times Pb\_emod + 0.0372 \tag{4.1}$$

The unit of tensile elastic modulus (E_t) in the formula is GPa. After substituting the tensile modulus (9.615 GPa) into Equation (4.1), the value of the bond effective modulus can be obtained, and it was 7.35 GPa.

4.2.2.3   Calibration of the effective modulus

The value of the bond effective modulus was unchanged, and the effective modulus was calibrated by a biaxial compression numerical test. Note that the uniaxial compression test was simulated by the numerical biaxial test with low confining pressure (0.1 MPa). This was because the calculation of the lateral strain of the biaxial test was easy and the value was stable, and it was more convenient to get the Poisson's ratio. During the numerical test, the effective modulus (Emod) were changed to 0, 1 GPa, 5 GPa, 10 GPa, and 20 GPa. And the corresponding relationship between the effective modulus and the elastic modulus can be obtained by the fitting (Figure 4.5).

According to the experimental results of numerical simulation, the linear relationship between the effective modulus (Emod) and the compression modulus (E_c) is:

$$E_c = 0.4353 \times Emod + 7.105 \tag{4.2}$$

The unit of the compression modulus (E_c) in the formula is GPa. After substituting the tensile modulus (12.5 GPa) into Equation (4.2), the value of the effective modulus can be obtained, and it was 12.39 GPa.

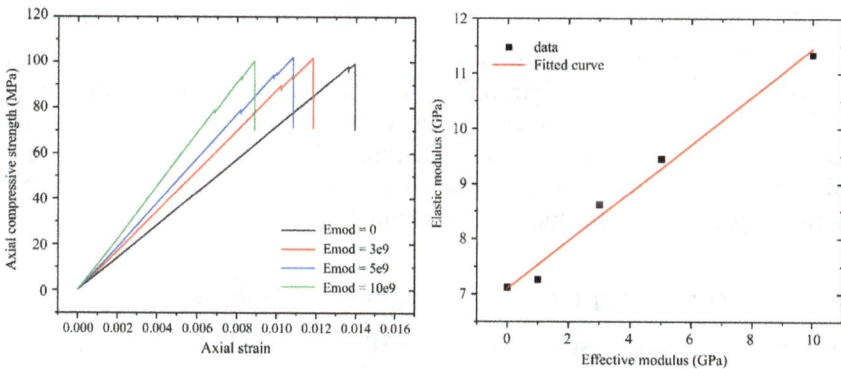

Figure 4.5 Compression test curves (left) and the parameter fitting (right) under different effective modulus.

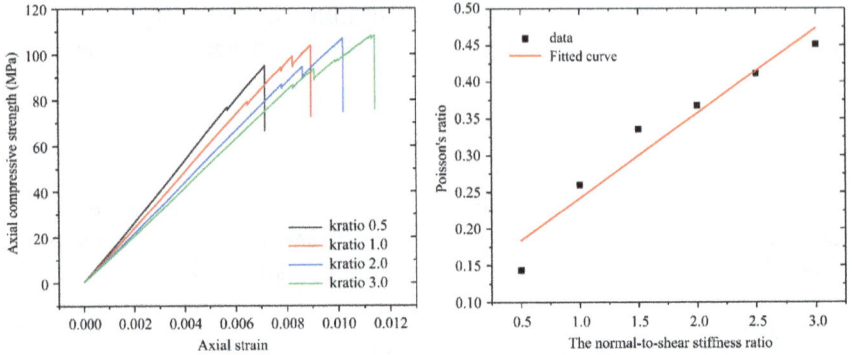

Figure 4.6 Compression test curves (left) and parameter fitting (right) under different normal-to-shear stiffness ratio.

4.2.2.4   Calibration of the normal-to-shear stiffness ratio

The value of the bond effective modulus and effective modulus were unchanged, and the normal-to-shear stiffness ratio (Kratio) was changed to 0.5, 1.0, 1.5, 2.0, and 3.0. Then the corresponding relationship between the Poisson's ratio and normal-to-shear stiffness ratio can be obtained by fitting (Figure 4.6).

According to the experimental results of numerical simulation, the linear relationship between Poisson's ratio (μ) and normal-to-shear stiffness ratio (Kratio) is:

$$\mu = 0.1155 \times \text{kratio} + 0.1261 \tag{4.3}$$

After substituting the Poisson's ratio (0.25) into Equation (4.3), the value of the normal-to-shear stiffness ratio can be obtained, and it was 1.07.

4.2.2.5   Calibration of the bond normal-to-shear stiffness ratio

The bond normal-to-shear stiffness ratio (k_{ns}) was the ratio of normal bond strength to tangential bond strength. The failure modes of the samples under different k_{ns} (Figure 4.7) indicated that when the value of k_{ns} was large, the sample was prone to shear failure; when the value of k_{ns} was small, the sample was likely to brittle failure. Therefore, according to rock failure in the actual experiment, the value was determined to be 1.2.

4.2.2.6   Calibration of the tensile strength and cohesion strength

Assuming that the cohesion strength was 10 MPa, based on the bond normal-to-shear stiffness ratio ($k_{ns} = 1.2$), the tensile strength was 12 MPa. And the value of these two strengths was set as the base bond strength. Then the base

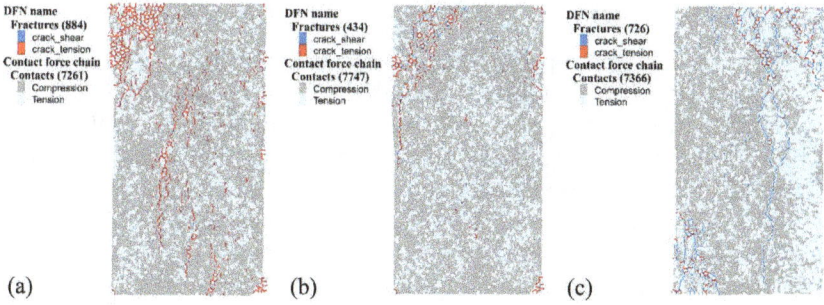

Figure 4.7 Comparison of compression test under different bond normal-to-shear stiffness ratio. (a) $k_{ns} = 0.5$, (b) $k_{ns} = 1.0$, (c) $k_{ns} = 2.0$.

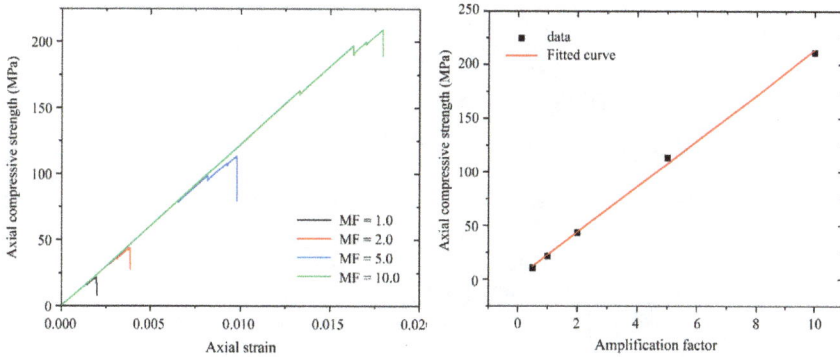

Figure 4.8 Compression test curves (left) and parameter fitting (right) under different amplification factors of the microscopic bonding parameter.

bond strengths were multiplied by coefficients of 0.5, 1.0, 2.0, 5.0, and 10.0 to obtain the peak strengths. This coefficient was defined as the amplification factor of the microscopic bonding parameter (MF).

According to the experimental results of numerical simulation (Figure 4.8), the linear relationship between axial compressive strength (σ_c) and the amplification factor (MF) is:

$$\sigma_c = 21.214 \text{ MR} + 1.6254 \tag{4.4}$$

The unit of axial compressive strength (σ_c) in the formula is MPa. After substituting the compressive strength (55 MPa) into Equation (4.4), the value of the amplification factor can be obtained, and it was 2.516. Therefore, the cohesion strength was 26.16 MPa and the tensile strength was 30.19 MPa.

4.2.2.7 Further adjustment of parameters and staged simulation

According to the micro-parameters determined earlier, compression tests and tensile tests were conducted to analyze the mechanism of macro settings under corresponding micro-parameters. Due to the mutual influence of each parameter, it was required to further modify the contact model's micro-parameters. Besides, in the initial compaction stage (strain less than 0.002), the initial elastic modulus was about 10% of the linear stage's elastic modulus. This feature was simulated through the programming of the PFC. The numerical simulation experiments of the remaining four mechanical states were also carried out through the same parameter calibration method. The relevant numerical results are shown in Figure 4.9.

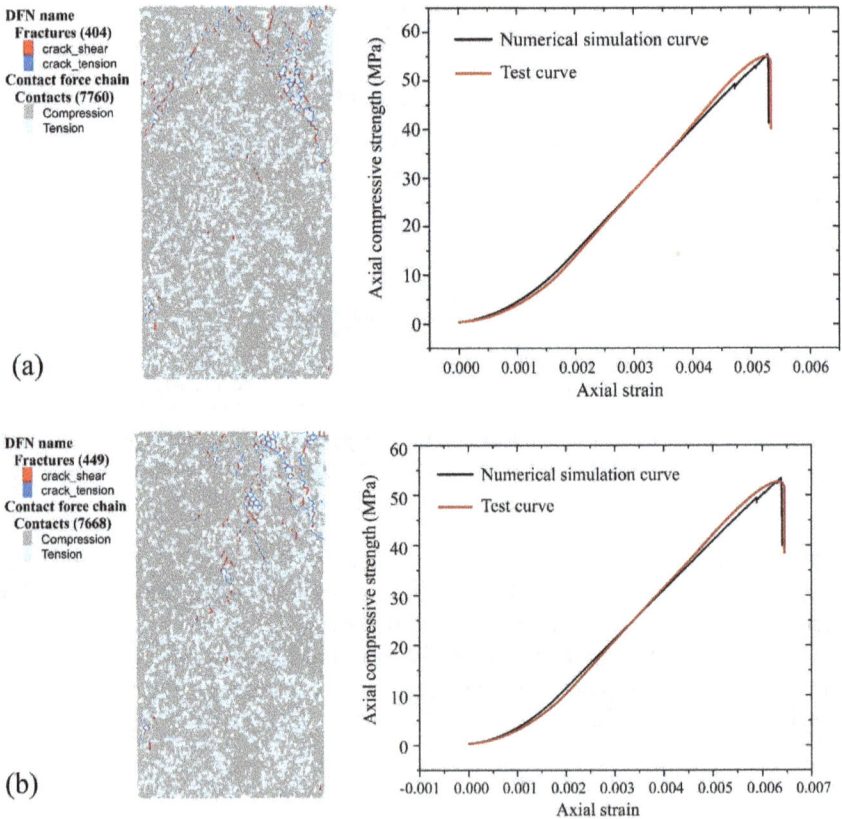

Figure 4.9 Comparison of numerical simulation results: (a) uniaxial compression in the dry state; (b) uniaxial compression in the saturation state; (c) hydraulic coupling under confining pressure 0.3 MPa; (d) hydraulic coupling under confining pressure 1 MPa; (e) hydraulic coupling under confining pressure 3 MPa.

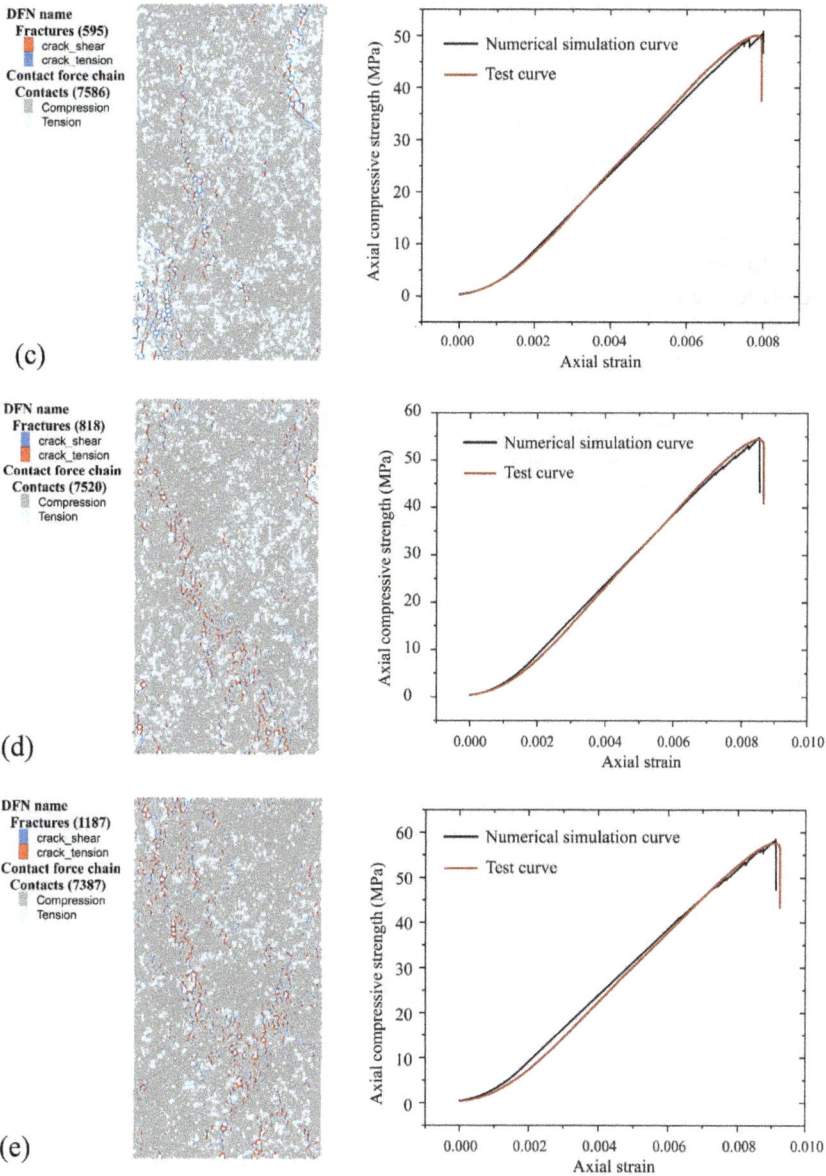

(c)

(d)

(e)

Figure 4.9 (Continued)

It can be found from the comparison that the numerical experiment results of the PFC were in good agreement with the experimental data, and the corresponding law of the mechanical properties of the sample can be realized from a microscopic perspective. This method could provide a data basis for the quantitative analysis of the characteristics under different mechanical states.

4.2.3 Quantitative analysis of mechanical properties of rock mass

The relevant mechanical parameters involved in the numerical simulation process are shown in Table 4.1. The trend of parameters related to strength and deformation is shown in Figure 4.10.

During the numerical experiment, six key micro-parameters were used to determine the mechanical properties of the rock mass, namely the effective modulus, bond effective modulus, normal-to-shear stiffness ratio, bond normal-to-shear stiffness ratio, tensile strength, and cohesion strength. Combining the constitutive relationship of the parallel bond model and the sensitivity analysis of critical micro-parameters, the mechanical properties in different states can be summarized as follows.

As the rock mass changed from the dry state to the saturated state and then to the hydraulic coupling state, the elastic modulus continued to decrease,

Table 4.1 Parameters under different mechanical states (obtained by the numerical simulation)

Parameters	Uniaxial compression in dry state	Uniaxial compression in saturation state	Hydraulic coupling (0.3 MPa)	Hydraulic coupling (1 MPa)	Hydraulic coupling (3 MPa)
Peak intensity (MPa)/ test data	54.9	52.1	50.2	54.5	58.6
Poisson's ratio/test data	0.25	0.25	0.26	0.26	0.26
Elastic modulus (GPa)/test data	12.5	9.82	7.5	7.42	7.21
Tensile modulus (GPa)/theoretical data	9.615	7.262	5.762	5.707	5.54
Effective modulus (GPa)	6.518	5.112	2.845	2.78	1.532
Bond effective modulus (GPa)	6.615	4.852	4.373	4.351	4.223
Normal-to-shear stiffness ratio	1.2	1.05	1.0	1.0	1.0
Bond normal-to-shear stiffness ratio	1.2	1.2	1.2	1.2	1.2
Tensile strength (MPa)	28.49	28.2	27.3	23.39	21.89
Cohesion strength (MPa)	23.74	23.5	22.8	19.99	18.24

Source: Zou et al. (2022)

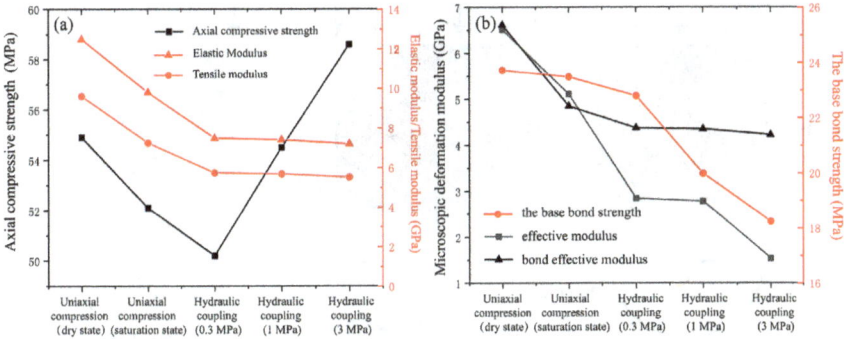

Figure 4.10 The trend of parameters related to strength and deformation: (a) comparison of macro-parameters; (b) comparison of micro-parameters.

Source: Zou et al. (2022)

while the plastic characteristics continued to strengthen. Microscopically, the effective modulus and bond effective modulus between the particles were reduced. That was to say, when the participation degree of water was getting higher, the bond effective modulus of the particles was reduced during the compression and tension. After the bond contact between the particles broke and the parallel bond model degenerated into linear contact, the effective modulus of the particles was also reduced. The participation of water accelerated the deformation tendency between particles and eventually led to an increase in the plastic deformation of the rock mass.

Similarly, as the participation degree of water was higher, the bond contact between particles would be smaller. Furthermore, the number of fractures between particles increased, which accelerated the destruction of rock mass and eventually led to a decrease in peak strength.

In the saturated state, when the space occupied by the water was compressed, the water would be transported, which resulted in the limited impact of water on the particle contact. In the hydraulic coupling state, the storage space's compression did not affect the water pressure which could continue to weaken the contact bond of the particles. The continuous existence of water pressure would accelerate the breaking of the contact bond between the particles. And then the water pressure would further promote the extension of these microcracks. This was the reason for the difference in mechanical properties between the saturated state and the hydraulic coupling state.

When the water pressure increased from 0.3 MPa to 1 MPa and then increased to 3 MPa, the micro-mechanical parameters continued to decrease, even though the macro-mechanical strength had increased. This phenomenon was because the increase in water pressure restricted the rock mass's deformation to a certain extent and further weakened the bond contact

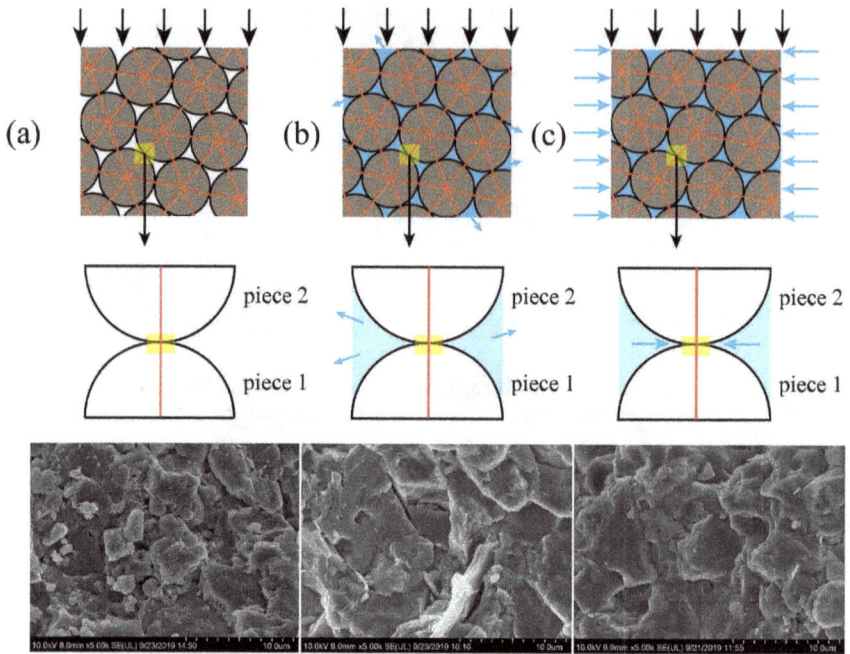

Figure 4.11 The particles under different mechanical states. (a) Uniaxial compression in the dry state. (b) Uniaxial compression in saturation state. (c) Hydraulic coupling with confining pressure of 1 MPa.

Source: Zou et al. (2022)

between the particles. It should be noted that the reduction in elastic modulus on macro-parameters was relatively small. When the contact broke into the microcracks, the increase of water pressure had little effect on the bond effective modulus between particles but a more significant impact on the effective modulus.

The results of SEM under different mechanical states were selected for comparative analysis (Figure 4.11). Note that the chosen micrographs here were all taken from the fractured surface of the rock mass. Observation of the microstructure showed that the fracture surface in the dry state was rough with more loose material. The loose content in the hydraulic coupling state was the least, followed by the saturated state. When the degree of water participation was higher, the energy for breaking the bond between the particles was lower; the fractured surface was relatively smooth; and fewer free particles were generated. This phenomenon not only proved that it was feasible to establish a micro-to-macro link through numerical simulation but also provided evidence for the special mechanical state of hydraulic coupling.

4.3   Analysis of the evolution process of rock mass

According to the previous analysis, the dangerous rock masses on the reservoir bank were in the mechanical environment with alternating cycles of dry state and hydraulic coupling state. Therefore, the uniaxial compression (dry state) and hydraulic coupling state (0.3 MPa) were selected to compare and analyze in this section. During the experimental process, the energy release process was recorded by the acoustic emission experiment, and the rock deformation process was analyzed by the non-contact full-field strain measurement system (VIC-3D). On this basis, the entire evolution process of the rock mass was investigated.

4.3.1   Evolution process of deformation

The non-contact full-field strain measurement system (VIC-3D) was used to analyze the deformation of the sample. Digital image correlation and tracking is an optical method that employs tracking and image registration techniques for accurate 2D and 3D measurements of changes in images. This method is often used to measure full-field displacement and strains and is widely applied in many areas of science and engineering. Compared to strain gauges and extensometers, the amount of information gathered about the fine details of deformation during mechanical tests is increased due to the ability to provide both local and average data using digital image correlation (Zhou et al., 2018).

VIC-3D system includes a calibration board, light source, a high-speed camera, and processing software. When measuring the deformation by VIC-3D, there are mainly four steps: spray test subject, calibrate, record event, and calculation deformation. The relevant test steps are shown in Figure 4.12.

During the test, the picture data was recorded every 0.5 s. According to the data processing program built-in VIC-3D, the results of the image were calculated and analyzed, and the deformation evolution process of the sample can be obtained. The experimental results of uniaxial compression (dry state) and hydraulic coupling (0.3 MPa) are shown in Figures 4.13 and 4.14.

Besides, through further extraction of the data, the evolution process of the average principal strain under different mechanical states can be obtained (Figure 4.15).

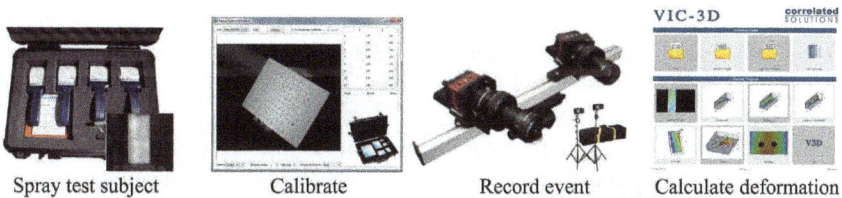

Spray test subject   Calibrate   Record event   Calculate deformation

Figure 4.12 Test method of VIC-3D system.

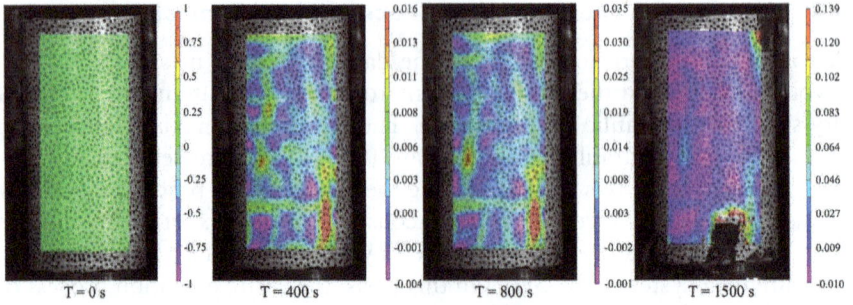

Figure 4.13 Evolution process of principal strain under uniaxial compression (dry state).

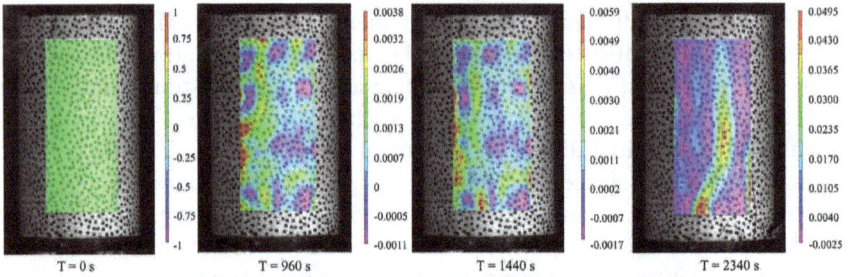

Figure 4.14 Evolution process of principal strain under hydraulic coupling (0.3 MPa).

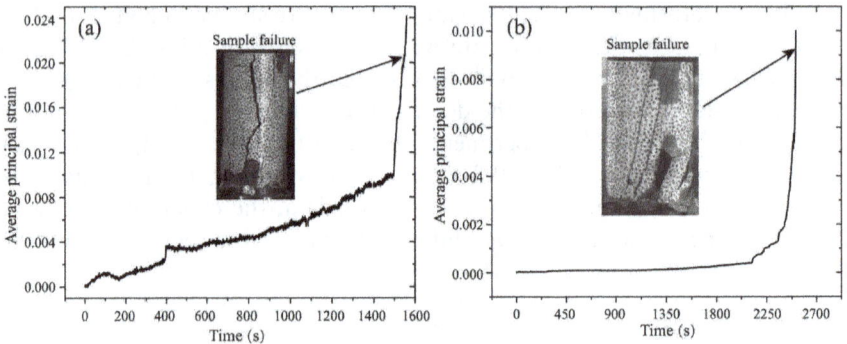

Figure 4.15 Evolution process of average principal strain: (a) uniaxial compression in the dry state; (b) hydraulic coupling (0.3 MPa).

Analysis of the deformation evolution process indicated that the deformation trend of the uniaxial compression state and hydraulic coupling state was similar. Specifically, the amount of deformation rate in the early stage was relatively gentle. During the elastic deformation stage, the deformation rate

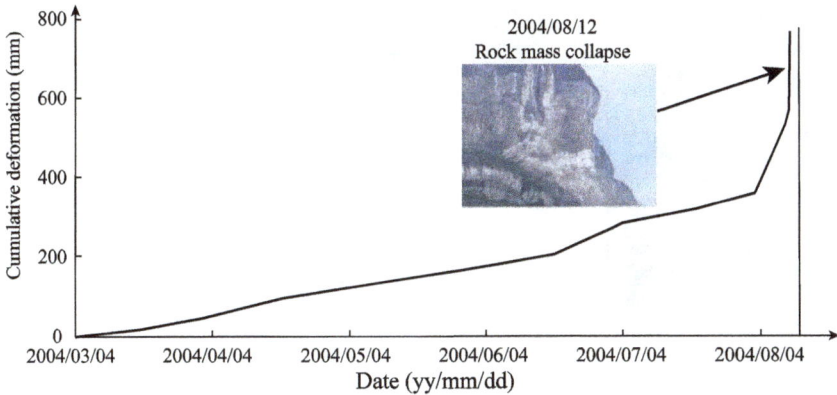

Figure 4.16 Deformation monitoring curve before Zengziyan collapse.

was unchanged. After entering the acceleration stage, the deformation rate increased rapidly, eventually leading to the failure of the sample.

The Zengziyan collapse, a typical steep dangerous rock mass, can provide relevant reference monitoring data for the deformation trend (Figure 4.16; He et al., 2019a). The monitoring data showed that the deformation rate of Zengziyan rock increased slowly in four months before the collapse, and the deformation rate increased rapidly ten days before the collapse. This displacement trend was similar to the deformation evolution process obtained from the experiment, which proved that the compression experiment of low confining pressure can adequately reflect the evolution process of the steep dangerous rock mass.

In the next section, the acoustic emission experimental data would be used to analyze the evolution process quantitatively.

Based on the generalized model of mechanics and the previous parameter calibration results, the failure analyses under uniaxial compression and hydraulic coupling were carried out. A simplified model involving a hard rock mass in the upper part and a weak rock mass in the lower part was used in the numerical simulation. It should be noted that the focus of the research in this section was to analyze its conceptual structural model qualitatively, and the numerical simulations carried out for the JDRM will be explained in detail in Chapter 6.

The numerical model and the corresponding stress–strain curves are shown in Figure 4.17. And the parameters of the hard rock used here are shown in Table 4.2. The mechanical parameters of the soft rock located in the lower part can refer to in Table 4.1.

Through the analysis and comparison of the failure models (Figure 4.18), it can be found that since the strength of the base area was low, during the continuous deformation process, the accumulation of deformation of the

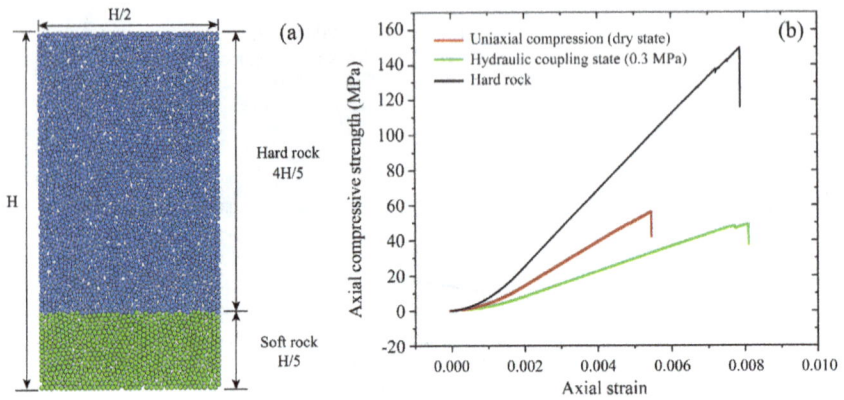

Figure 4.17 The numerical model (a) and the corresponding stress–strain curves (b).

Table 4.2 The parameters of the hard rock

Effective modulus (GPa)	Bond effective modulus (GPa)	Normal-to-shear stiffness ratio	Bond normal-to-shear stiffness ratio	Tensile strength (MPa)
11.2	11.25	1.2	1.2	77.69
Cohesion (MPa)	Peak intensity (MPa)	Poisson's ratio	Elastic modulus (GPa)	Tensile modulus (GPa)
64.74	150	0.25	21.5	16.5

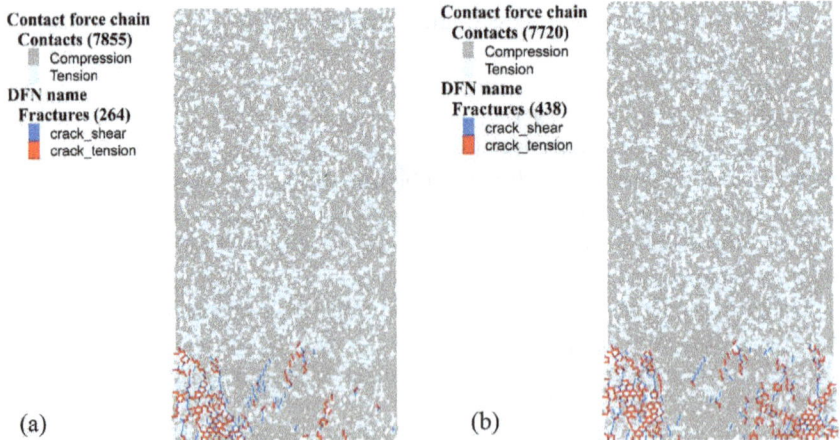

Figure 4.18 Failure models of uniaxial compression (a) and hydraulic coupling (b).

base rock mass became the critical factor to control the failure of the numerical models. This failure mode was consistent with the mechanical model proposed in Chapter 2. Under the same mechanical environment, if the base area's strength were lower, more cracks would be generated. Furthermore, as higher uneven stress in the base area is triggered by more cracks, the rock mass may disintegrate into more fragments during the failure process.

4.3.2 Damage evolution trend based on acoustic emission experiment

If the rock is deformed by a load, the stress distribution around a crack would be concentrated, and the strain energy would be high. When the strain energy is increased, micro-deformation occurs, cracks expand, strain energy works, energy is quickly released, and part of the stored energy produces transient elastic waves that are an acoustic emission phenomenon of rock materials. The equipment of acoustic emission can monitor the change of internal structural state during rock rupture by observing and analyzing this transient elastic wave. After receiving the acoustic waves in the micro-fracture inside the rock, the time, position, and intensity of the micro-fracture are recorded, and the mechanical behavior of rock failure and the current state of internal defects of rock materials can be analyzed (Zhou et al., 2010; Xu et al., 2016). Although there are extensive applications of acoustic emission technology in geotechnical engineering, this book is the first to apply acoustic emission testing technology directly to the simplified mechanical model of the submerged steep dangerous rock mass.

The acoustic emission signals involve position, amplitude, energy, and ring count that can be used to analyze the damage evolution during rock failure. In this book, the damage constitutive model was constructed by extracting the number of ringing from the acoustic emission signal (He, 2015).

A damage constitutive model can be constructed from an acoustic emission test. The damage variables are defined as follows:

$$D = \frac{A_d}{A},\tag{4.5}$$

where A_d is the damage sectional area which will increase with the damage process; A is the complete sectional area of the material in the non-damaged state. The ringing count of acoustic emission can characterize the damage state of the material. If section area A is completely destroyed because of damage, the cumulative acoustic emission count of the rock at this time is set to C_0, and the acoustic emission ringing count per unit area of the micro-element C_w is:

$$C_w = \frac{C_0}{A}.\tag{4.6}$$

Then, when the damage area of the section reaches A_d, the corresponding cumulative acoustic emission ringing count C_d is:

$$C_d = C_w A_d = \frac{C_0}{A} A_d. \tag{4.7}$$

Using the results from Equation (4.5) and Equation (4.7), the damage variable D can be obtained:

$$D = \frac{C_d}{C_0}. \tag{4.8}$$

During the test, because of the different conditions of the rock sample rupture and other factors, the rock damage state and the damage state may not correspond, so the damage variable is corrected to:

$$D = D_u \frac{C_d}{C_0}. \tag{4.9}$$

In Equation (4.9), the D_u is the damage threshold that can be defined by the following equation:

$$D_u = 1 - \frac{\sigma_r}{\sigma_p}, \tag{4.10}$$

where σ_p is the peak strength of the rock sample and σ_r is the residual strength of the rock sample. When σ_r and σ_p are equal, D_u is equal to 0, the rock sample can be regarded as an ideal elastoplastic material, and the damage variable is 0; when σ_r is equal to 0, D_u is equal to 1, which indicates that the rock would fail during compression.

Experimental results of the stress–strain curve and the ringing count distribution are shown in Figure 4.19. The strength of the sample in a uniaxial compression was 54.9 MPa, which was 8.56% higher than the 50.2 MPa of the sample under hydraulic coupling. Since there was no lateral constraint during uniaxial compression, the residual strength in this mechanical state was negligible. And the residual strength under the hydraulic coupling test was 4.42 MPa. After the time changed from the elastic phase to the plastic phase, the ringing count began to gradually increase, and the ringing count reached its peak at the moment the sample fails. Specifically, the peak value of the uniaxial compression test was much larger than the ringing count under the hydraulic coupling. The maximum value of the ringing count under the uniaxial compression test was much larger than the maximum value under hydraulic coupling.

According to the aforementioned calculation equation, the D_u of hydraulic coupling triaxial compression was 0.912, and the D_u of uniaxial compression was 1, which indicated the ultimate damage of sample under uniaxial

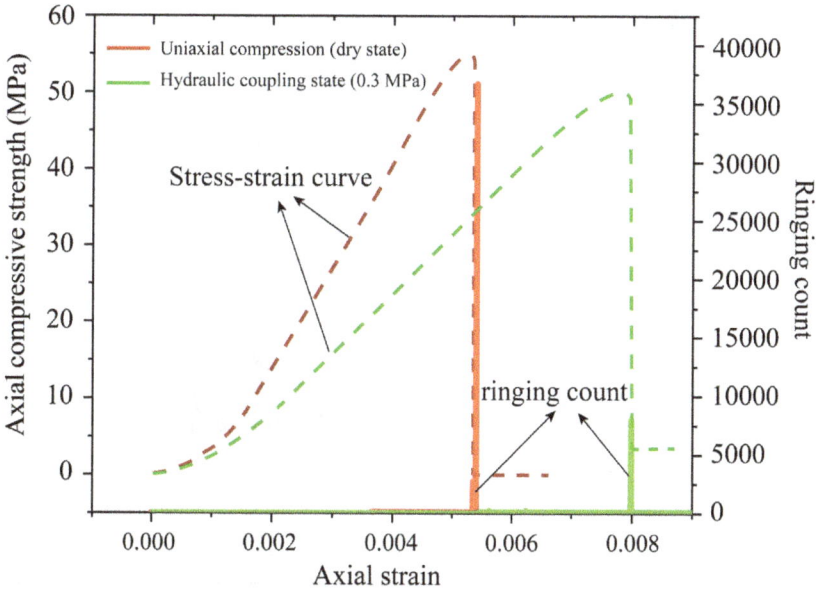

Figure 4.19 Axial pressure curve and ringing count distribution.

compression test was greater than the hydraulic coupling. The equation of the damage variable under hydraulic coupling is:

$$D = 0.912 \frac{C_d}{C_0}. \tag{4.11}$$

Damage variables under the uniaxial compression is:

$$D = \frac{C_d}{C_0}. \tag{4.12}$$

Based on the experimental data, the damage variable curves obtained by Equations (4.11) and (4.12) are shown in Figure 4.20.

By analyzing Figure 4.20, the damage variable curves can be divided into two stages: the stable phase and the rising phase. The stable phase can also be defined as the initial damage phase. At this stage, the damage variable was zero or close to zero. There was only slight coordinated deformation or no deformation for initial microcracks and micropores because the sample was in the elastic deformation stage, and there was not enough energy to cause further deformation. Therefore, no new micro-fractures occurred, and the ringing count was rare. The damage variable under the hydraulic coupling at this stage was similar to the uniaxial compression test.

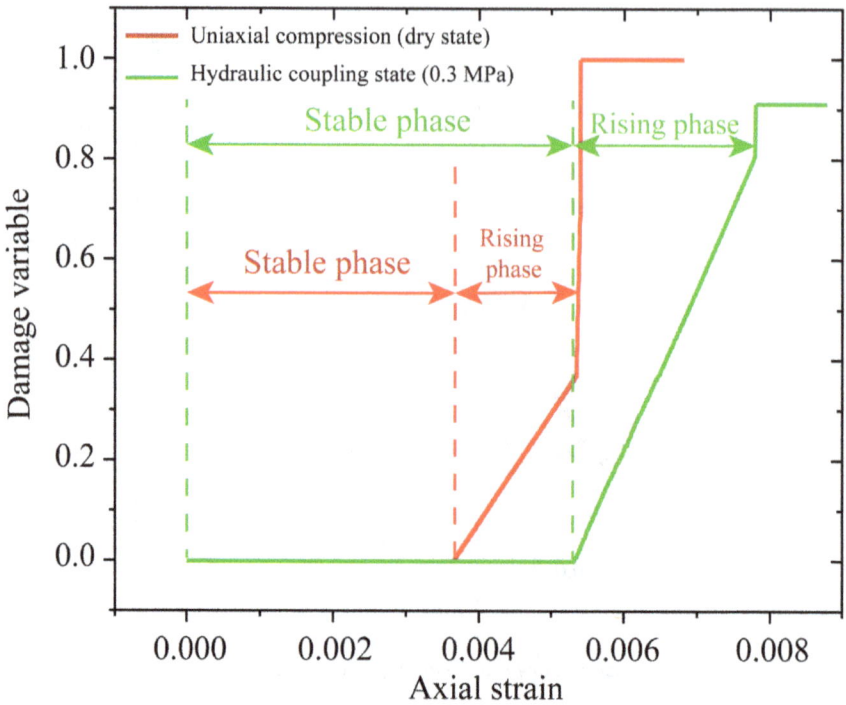

Figure 4.20 Damage variable curve.

The second phase was the rising phase, which can also be defined as the damage development and mutation phase. The damage variable varied greatly during this phase. As the axial pressure increased, the sample entered the plastic phase, and the strain energy began to expand the microcracks and micropores. The new micro-fractures gradually occurred, and the acoustic emission activity was gradually active. At the beginning of the stage, the damage variable continuously increased. However, at the end of the stage, the bearing capacity of the sample reached the limit, the curve rapidly increased, and the damage variable abruptly increased. As the microcracks expanded, fused, and penetrated, the sample underwent macroscopic damage, and the acoustic emission activity reached its peak at the end of this stage.

In the rising phase, the uniaxial compression test was substantially different from the hydraulic coupling test. In the hydraulic coupling test, since the water was in direct contact with the rock mass, the water pressure directly participated in the expansion of the fractures. With the extension of the original cracks and the continuous generation of new cracks, under the action of water pressure, the water will track the crack, and every mineral particle on all free surfaces would be wet, thereby weakening the intergranular relationship, reducing the strength, and then speeded up the crack propagation.

Therefore, the damage accumulation in the rising phase was more thorough and effective, the increasing rate was higher and the time lasted longer during the rising phase. Furthermore, because of the accumulation of early damage development, at the end of the rising phase, the sudden increase in the number of the ringing count was smaller than the uniaxial compression test.

Residual strength after the peak intensity normally represented the third stage and was known as the post-peak rising stage (Arash et al., 2014; He, 2015; Li et al., 2017b). However, for the base rock mass, the residual strength of rock mass under low confining pressure was small, so the dangerous rock mass would be directly unstable after reaching the peak intensity. Therefore, only the stable phase and rising phase were analyzed, and the acoustic emission signals generated after the peak strength were no longer considered in this book.

When the damage variable-strain and stress–strain curves under hydraulic coupling action were superimposed (Figure 4.21), it can be found that the correspondence between different stages was good. The trend of damage variable with strain can characterize the damage evolution process of a rock. In the elastic stage, the damage variable was in the stable phase; after entering the plastic stage, as the rock deformation increased, the damage variable also increased gradually; when the stress reached the peak value, the damage evolution curve abruptly increased and the rock sample broke. It should be noted that rock damage was an irreversible process, thus the damage variable

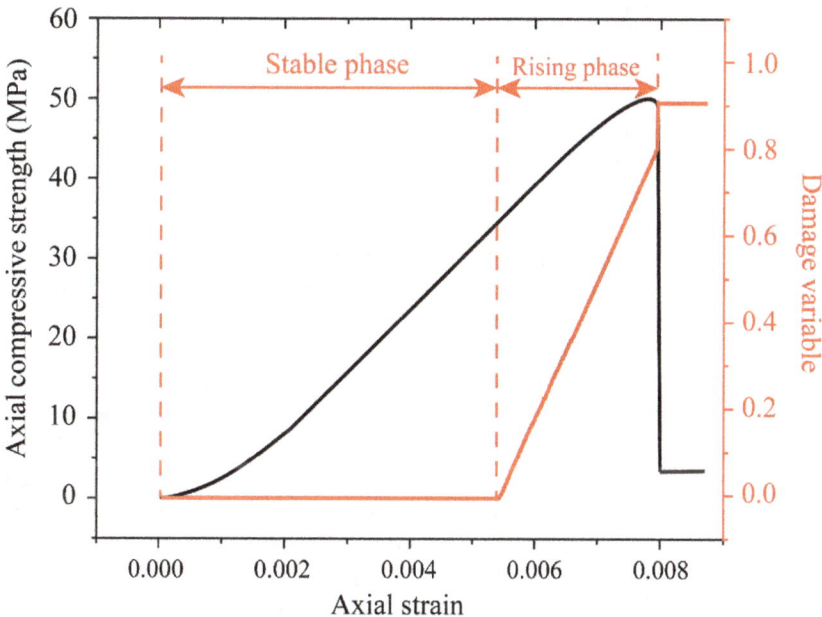

Figure 4.21 Damage variable and stress–strain curve.

can only increase with time or with strain and cannot be reduced. In addition, the damage evolution of the sample was a complex nonlinear acceleration process that was consistent with the characteristics exhibited in the actual damage process.

4.3.3 The failure mechanism by introducing damage evolution

Combined with the actual working conditions, the damage evolution was introduced into the mechanical model to analyze the failure mechanism of the submerged dangerous rock mass. According to the mechanical environment of steep dangerous rock mass, the effective bearing area based on damage evolution can be obtained as follows:

$$A_D = A_0 (1-D),$$
(4.13)

where A_0 is the effective bearing area of the rock mass when the rock is not damaged or the initial damage degree is very small, and A_D is the effective bearing area in the damage process which decreases significantly with the evolution of the damage.

The effective stress experienced by the rock mass in the damage zone is as follows:

$$\sigma_D = \frac{\gamma(H-h)A_0}{A_D} = \frac{\gamma(H-h)}{1-D}.$$
(4.14)

The dangerous rock mass would collapse when the effective stress σ_D is greater than the compressive strength σ_c. Since the axial pressure is the self-weight of the upper rock mass, the value of the axial pressure is constant. Furthermore, the effective axial pressure continues to increase while the damage accumulates.

When the damage variable was introduced, the damage evolution of the submerged steep dangerous rock mass in the reservoir area can be divided into three stages (Figure 4.22):

(1) Formation of the bottom damage zone. The degradation zone was formed by the periodic fluctuation of the water level in the reservoir area. The bottom damage zone was developed by the base rock mass located in the degradation zone. For the submerged steep dangerous rock mass, the degradation of the base rock mass was the key factor to promote the instability of the dangerous rock mass.
(2) Damage evolution with progressive deformation. As the number of fluctuations in the water level increased, the strength of the rock mass continued to decrease, that was, the value of σ_c was reduced. While the degree of damage continued to increase (i.e., the value of D increases),

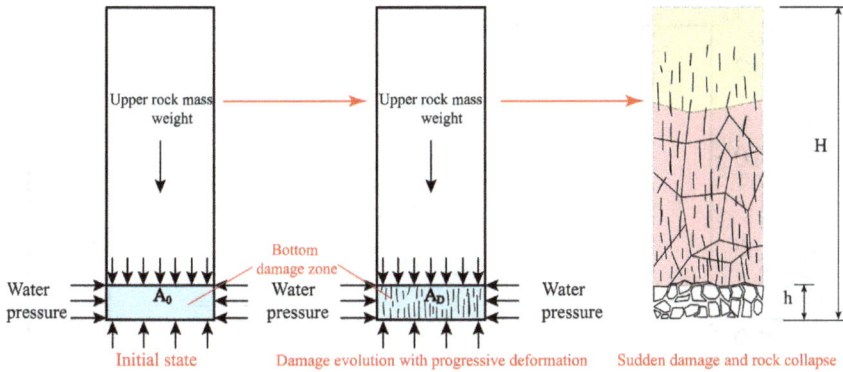

Figure 4.22 Damage evolution of submerged dangerous rock mass.

Source: Wang et al. (2020c)

the value of σ_D increased. Since the decrease of σ_c and the increase of σ_D were concurrent, the ratio of σ_D to σ_c would be smaller.

(3) Sudden damage and rock collapse. When the ratio of σ_D to σ_c was less than 1, the dangerous rock mass would be unstable.

It should be noted that the division of the evolution process mentioned here did not consider the time-dependent effect. The specific evolution process will be further discussed in the next chapter through the experiment.

4.3.4 Amplification of the effective axial pressure

The value of $1/(1-D)$ in Equation (4.14) can be defined as the value of the magnification of the effective axial pressure, and the introduction of the damage variable was used to quantify the amplification effect of the damage in the actual working conditions. As the amount of deformation increased, the amplification effect of the effective axial pressure because of damage is shown in Figure 4.23, which can be generalized to the whole process curve of the instability of the submerged steep dangerous rock mass. According to the analysis of damage variables, the phenomenon of progressive gradual deformation and sudden failure was prominent. In the stable phase, the strength reduction of rock mass under the reservoir water fluctuation was the key factor to the stability of the dangerous rock mass. Damage did not magnify the axial pressure of the upper rock mass. In the rising phase, under the unit deformation, the amplification effect caused by the damage was far greater than the speed reflected by the deformation. According to the test in this section, after the same deformation occurred, the cumulative amount of damage under hydraulic coupling was greater, and the presence of water accelerated the deterioration of the rock mass and crack propagation.

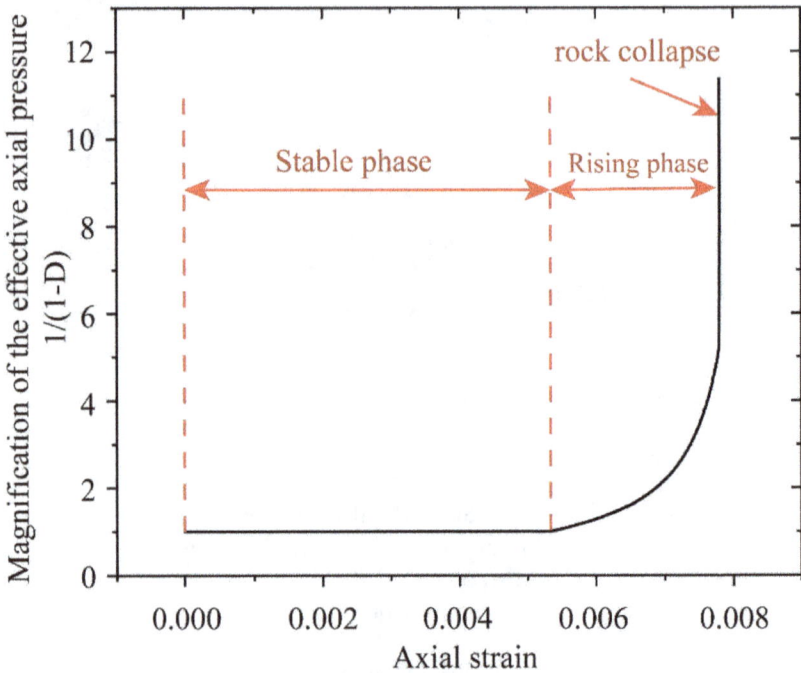

Figure 4.23 Amplification of the effective axial pressure.

Source: Wang et al. (2020c)

According to the monitoring data of the JDRM, the monitoring pressure of the base rock mass continued to increase. However, the self-weight of the overlying rock mass was constant and had always been 3.24 MPa. Continuous nonlinear increase of the base rock pressure proved the amplification effect of the damage. Furthermore, it showed that the current stage of JDRM was in the rising phase. Therefore, the trend of instability would continue to accelerate, and protection must be carried out as soon as possible.

4.4 Summary of this chapter

In this chapter, based on the actual mechanical state of the rocky reservoir bank, the particle flow code, digital image correlation method, and acoustic emission technology were used to study the evolution process under different mechanical states. Conclusions of this chapter are summarized as follows:

(1) As the rock mass changed from the dry state to the saturated state and then to the hydraulic coupling state, its strength continued to decrease, and the plastic characteristics continued to strengthen. The participation

of water reduced the bond contact and accelerated the deformation tendency between particles, eventually leading to an increase in plastic deformation and a decrease in peak strength.

(2) There were differences in mechanical properties of rock mass under the saturated state and hydraulic coupling state. In the hydraulic coupling state, the water would not be transferred due to the storage space compressed, and the water pressure can continue to weaken the contact bond of the particles. Therefore, the hydraulic coupling state can accelerate the breaking of the bond between the particles and further promote the extension of the microcracks.

(3) When the water pressure increased from 0.3 MPa to 1 MPa and then increased to 3 MPa, the increase in water pressure restricted the deformation of the rock mass to a certain extent and further weakened the bond between the particles, which led to an increase in peak intensity and a reduction in micro-parameters.

(4) According to the entire evolution process of rock mass investigated by the test, it can be found that the damage can magnify the effective stress experienced by the base rock mass. Moreover, incorporating the damage effects into the dangerous rock mass model created a nonlinear and accelerating deformation trend that accurately portrayed the failure of dangerous rock masses in the Three Gorges Reservoir area.

Chapter 5

Experimental research on the fluctuation of reservoir water level

This chapter studied the influence of reservoir water on the evolution of the submerged dangerous rock mass. The changing trend of mechanical parameters was analyzed through the weakening experiments under dry–wet cycles and numerical simulation. Afterward, the influence of the self-weight of upper rock on reservoir bank was examined through the transition of mechanical state test under different axial pressures. Finally, the evolution process considering the time-dependent effect was investigated by the creep experiments under the mechanical state transition. The research results obtained in this chapter were of great significance for studying the evolution characteristics of the dangerous rock mass on the reservoir bank.

5.1 The weakening experiments under dry–wet cycles

In this section, the weakening experiments under dry–wet cycles were used to study the deterioration trend of rock masses. For relevant experimental methods, refer to Section 2.3. After 5 and 10 dry–wet cycles, the rock samples were conducted uniaxial compression test under the dry state and hydraulic coupling test under 0.3 MPa confining pressure, respectively. The results of the weakening experiments under dry–wet cycles are shown in Figure 5.1.

Using the method stated in Chapter 4, the evolution trend of microscopic parameters under dry–wet cycles was also discussed in this section. The macro-micro-parameters of the rock mass under dry–wet cycles are shown in Table 5.1. And the numerical simulation results under the dry–wet cycles are shown in Figure 5.2.

According to Figures 5.1 and 5.2, it can be found that the theoretical curves obtained by numerical simulation matched well with the experimental curves. Furthermore, the quantitative relationship between macro-parameters and micro-parameters can be realized, which provided a data basis for the influence of dry–wet cycles on the mechanical properties of the rock mass. The trends of mechanical parameters of rock mass under the dry–wet cycle test are shown in Figure 5.3.

DOI: 10.1201/9781003347163-5

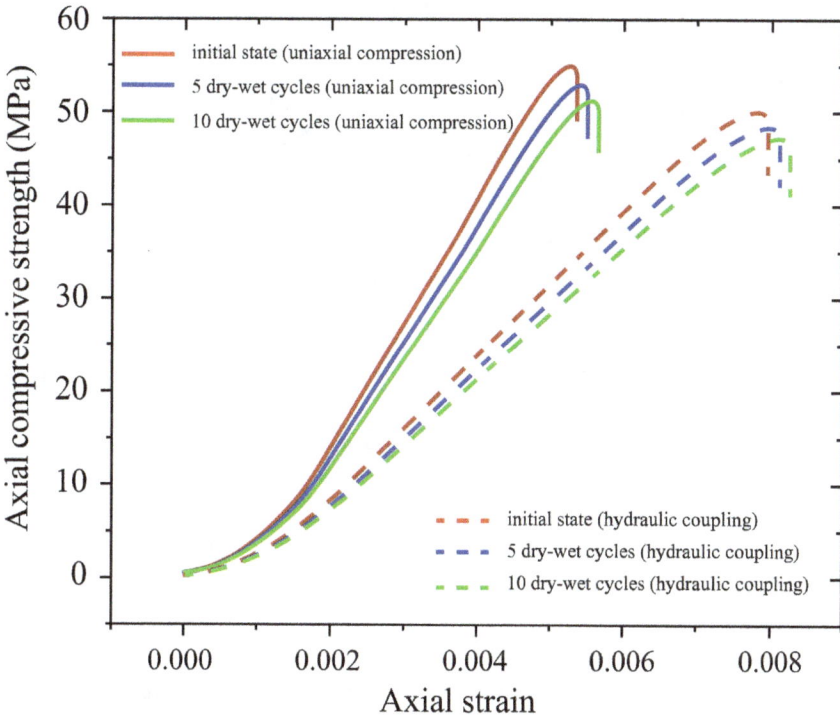

Figure 5.1 The weakening experiments under dry–wet cycles.

Table 5.1 Macro-micro parameters of rock mass under dry–wet cycles

Parameters	Uniaxial compression (dry state)			Hydraulic coupling (0.3 MPa)		
Dry–wet cycles	0	5	10	0	5	10
Peak intensity (MPa)	54.9	53.1	51.2	50.2	48	47.1
Poisson's ratio	0.25	0.25	0.25	0.26	0.26	0.26
Elastic modulus (GPa)	12.5	12.0	11.5	7.5	7.0	6.5
Tensile modulus (GPa)	9.615	9.23	8.84	5.762	5.38	5.02
Effective modulus (GPa)	6.518	6.257	5.996	2.845	2.655	2.466
Bond effective modulus (GPa)	6.615	6.35	6.08	4.373	4.1	3.82
Normal-to-shear stiffness ratio	1.2	1.2	1.2	1.0	1.0	1.0
Bond normal-to-shear stiffness ratio	1.2	1.2	1.2	1.2	1.2	1.2
Tensile strength (MPa)	2.849	2.755	2.657	2.386	2.271	2.229
Cohesion (MPa)	2.374	2.296	2.214	1.988	1.893	1.857

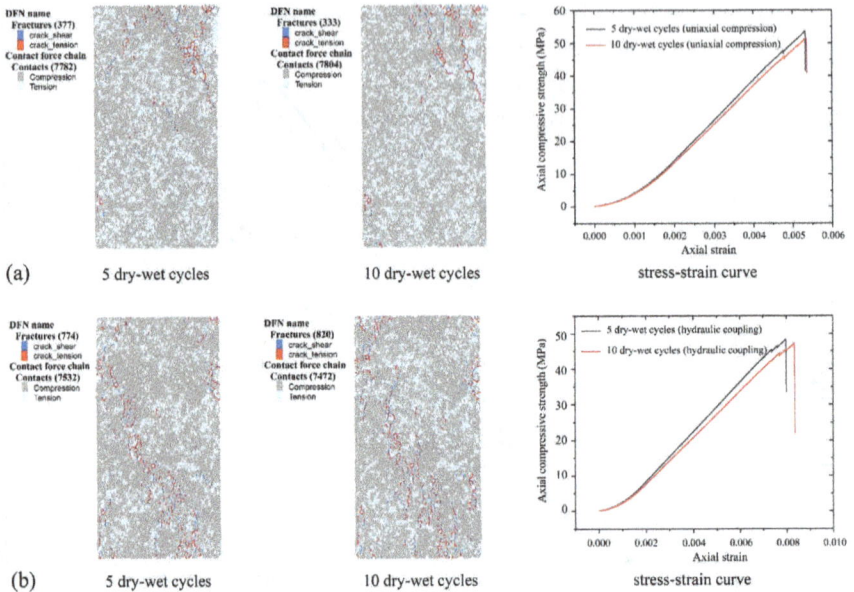

Figure 5.2 Numerical simulation of rock under the dry–wet cycles. (a) Numerical simulation of uniaxial compression under the dry–wet cycles. (b) Numerical simulation of the hydraulic coupling under the dry–wet cycles.

From the macro-parameters perspective, compared with the uniaxial compression test, the strength of the sample under hydraulic coupling was reduced by 8.5%. As the number of dry–wet cycles increased, the strength of the sample gradually decreased. The strength of the sample decreased by about 6% after ten dry–wet cycles in both the uniaxial compression state and the hydraulic coupling state. In this set of experiments, by comparing the peak strength of the initial uniaxial compression state (54.9 MPa) with that of the hydraulically coupled state after ten dry–wet cycles (47.1 MPa), it can be found that the decrease in the strength of the sample could reach 14.2%. This comparison indicated that the limestone used in the experiment had a relatively high degree of compactness and the change in the dry–wet cycle state had a small effect on the strength of the rock mass. However, the deformation characteristics of rocks under different mechanical conditions were more obviously caused by the dry–wet cycle. Under complex hydraulic coupling, the plastic characteristics of the rock were visible, which was manifested by a higher peak deformation and a lower peak strength. Notably, the sample's elastic modulus under complex hydraulic coupling was reduced by 40% compared with the uniaxial compression test results. As the number of dry–wet cycles increased, the elastic modulus of the sample gradually decreased. And the decrease in elastic modulus under hydraulic coupling was more significant than uniaxial compression.

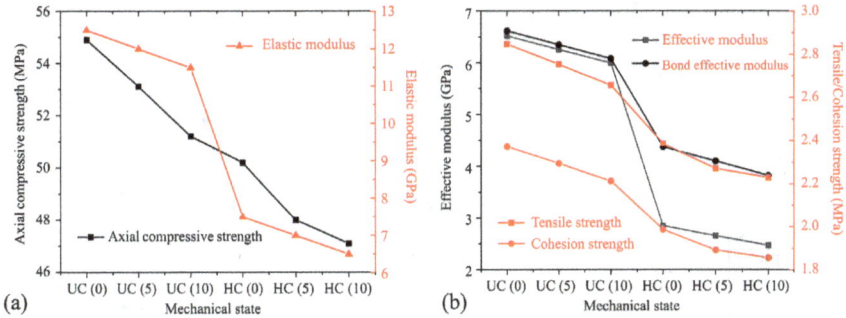

Figure 5.3 The trends of mechanical parameters of rock mass under the dry–wet cycle. UC is an abbreviation for uniaxial compression; HC is an abbreviation for hydraulic coupling; the number in brackets corresponds to the dry–wet cycles. (a) Macro-mechanical parameters. (b) Micromechanical parameters.

From the micro-parameters perspective, compared with the uniaxial compression test results, the force bond between the particles under complex hydraulic coupling was reduced with about 16%. As the number of dry–wet cycles increased, the force bond between the particles gradually decreased. In both the uniaxial compression state and the hydraulic coupling state, the force bond between the particles decreased by 6% after ten dry–wet cycles. It indicated that the force bonds between particles were more sensitive to changes in mechanical state. Similarly, the variation of the deformation parameters between the particles was more significant than the force bond. Among them, the effective modulus changed the most since it was controlled by the elastic modulus. And the effective modulus under complex hydraulic coupling was reduced with about 60% compared with the uniaxial compression test results.

According to the test results, it can be concluded that the dry–wet cycle had little effect on the dense limestone used in this book. However, when the mechanical state was cycled, the difference in rock plasticity and peak strength under different mechanical conditions became the critical factor in accelerating the deterioration of rock mass. Notably, the change in plastic state was greater than the peak intensity, which could cause a more significant impact on the mechanical properties of the rock mass when the mechanical state was changed.

5.2 Mechanical state transition test under continuous axial pressure

The influencing factor of the dry–wet cycle was considered in the weakening experiments. However, during the transition of the mechanical state, the rocky reservoir bank has always been bearing the weight of the overlying

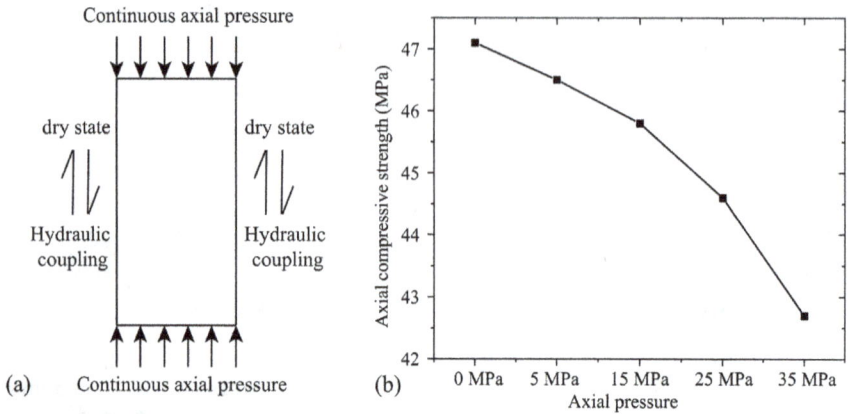

Figure 5.4 The mechanical state transition test under continuous axial pressure. (a) Mechanical environment. (b) The changing trend of axial pressure.

rock mass. In this section, the mechanical state transition tests under constant axial pressure were performed to study the deterioration trend of rock mass under complex mechanical conditions. The method of realizing the conversion of the mechanical environment under continuous axial pressure can refer to Chapter 3, and the related content will not be described in detail here.

Ten mechanical environment conversions were conducted under the axial pressures of 5 MPa, 15 MPa, 25 MPa, and 35 MPa, respectively (Figure 5.4a). During the tests, the next mechanical state will not be applied until the previous mechanical state's deformation remained stable. Therefore, there was sufficient time for the rock to complete the mechanical adjustment and reach the stress balance under each mechanical state. After completing the mechanical state transition, the uniaxial compression experiment was carried out in the dry state, and the peak intensity can be obtained (Figure 5.4b). The comparison results under different mechanical states are shown in Figure 5.5.

According to the test results, the following conclusions can be obtained:

(1) When considering both axial pressure (self-weight of upper rock mass) and mechanical state transition (the fluctuation of the reservoir water), the peak compressive strength of the dense limestone was lower than that under a single factor of dry–wet cycles.
(2) In the process of mechanical state transition, if the axial compression were higher, the ultimate compressive strength of the rock mass would be lower. In other words, if the weight of the overlying rock mass were greater, the influence of mechanical transition on the mechanical strength of the rock mass would be more significant. When the axial pressure was 35 MPa, the peak strength under the mechanical state transition was reduced by 22.23% compared with the initial state.

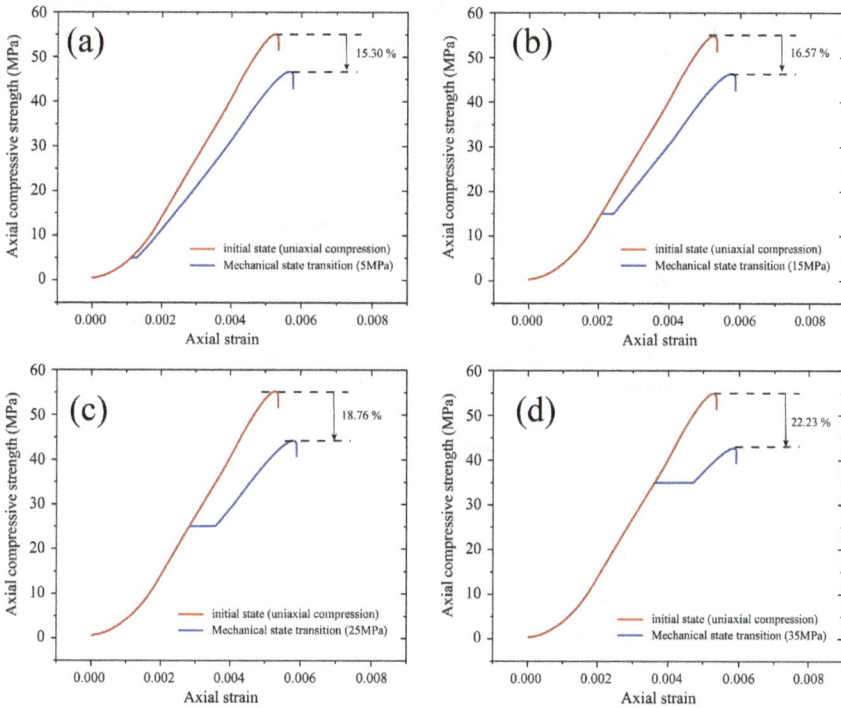

Figure 5.5 Comparison results under different mechanical states. (a) The axial pressure is 5 MPa. (b) The axial pressure is 15 MPa. (c) The axial pressure is 25 MPa. (d) The axial pressure is 35 MPa.

(3) Under certain axial pressure, the mechanical properties of the rock mass would be changed by the mechanical state transition. This change in mechanical properties was reflected in peak strength and creep deformation under the effect of axial pressure. And if the axial pressure were higher, the creep deformation would be more obvious.

Compared with the dry–wet cycle test without the axial pressure, the constant axial pressure caused initial damage to the rock. Furthermore, the mechanical state transition promoted the accumulation of damage, resulting in the continuous reduction of its mechanical properties. Besides, during the experiment in this section, there was sufficient time for the rock mass to complete the stress adjustment in a single mechanical state. This mechanical adjustment was embodied in creep deformation, which led to the accumulation of damage and a decrease in the peak intensity. When the weight of the overlying rock mass was higher, there was more rock damage in the base area, causing a more significant influence of the mechanical state transition on the rock mass.

5.3   Creep experiment under mechanical state transition

In this part of the experiment, three sets of comparative tests were conducted to study the influence of time-dependent effect, involving creep test under dry state, creep test under hydraulic coupling, and creep test with mechanical state transition. The comparison of these three creep conditions is shown in Figure 5.6. Based on the test results of the weakening experiments under dry–wet cycles, these three sets of creep tests were all loaded in units of 5 MPa. Notably, the creep experiment with the mechanical state transition was carried out by cycling between the dry state and the hydraulic coupling state. And each mechanical state transition was regarded as a hydrological cycle. Based on the on-site monitoring data, the pressure on the base of the rock mass reached its maximum value during the dry state of each hydrological cycle (in August each year). Therefore, the axial pressure will be increased in the dry state of the next cycle after completing a hydrological cycle.

5.3.1   Analysis of the test curves

The curves of the three sets of creep tests are shown in Figure 5.7, and the following conclusions can be drawn.

(1) The evolution process of these three creep tests was similar. Specifically, when the axial pressure did not exceed the long-term strength of the sample, the evolution curve underwent three stages under constant axial pressure: instantaneous elastic deformation, decelerated creep, and stable creep. When the axial pressure exceeded the long-term strength of the sample, the sample will enter the accelerated creep phase after undergoing the three phases mentioned previously, and it will eventually fail.

(2) The peak strengths obtained from the uniaxial compression creep tests and the hydraulic coupling creep tests were similar to the initial peak

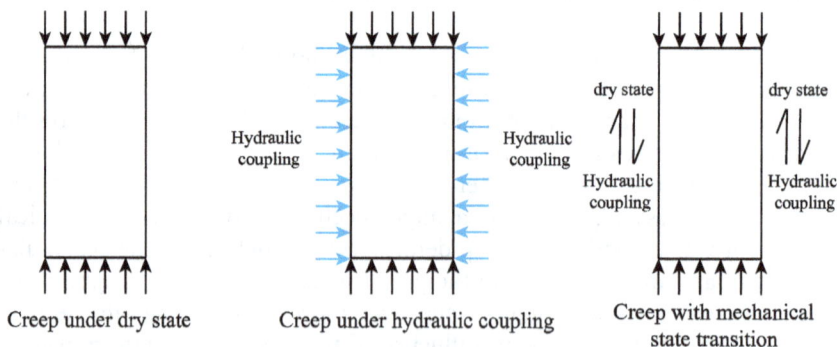

Figure 5.6 Comparison of the creep testing conditions.

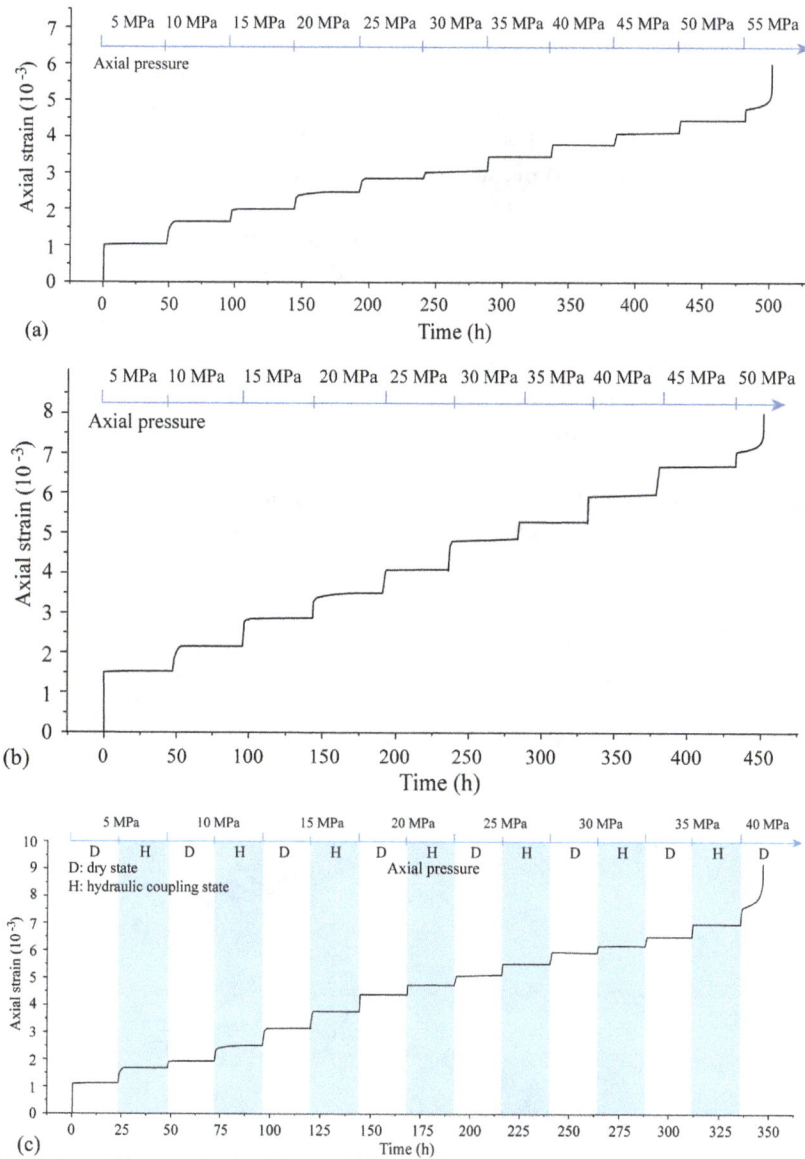

Figure 5.7 Results of the creep tests. (a) Creep experiment under uniaxial compression. (b) Creep experiment under hydraulic coupling. (c) Creep experiment with the mechanical state transition.

intensities obtained from the uniaxial compression tests and the hydraulic coupling tests, respectively. Therefore, the time-dependent effect caused by the creep tests in a single mechanical state on the compact limestone was small.

(3) For the creep experiment with mechanical state transition, the sample underwent seven mechanical state transitions before failure; its peak strength was reduced by 27.27% compared with the uniaxial compression creep test and by 20% compared with the hydraulic coupling creep test. This decrease in strength even exceeded the maximum decrease in strength of the sample after 10 dry–wet cycles (14.2%). Compared with the creep experiments in a single state, the transition of the mechanical state considering the time-dependent effect continued to promote the degradation of the rock mass, greatly reduced the peak strength of the rock mass, and accelerated the failure of the rock mass. In addition, when changing from the dry state to the hydraulic coupling state, the axial deformation of the sample increased. This was because there were certain differences in the mechanical characteristics of the sample under the different mechanical states, which was also demonstrated by the weakening experiments.

5.3.2 Analysis of the progressive cumulative deformation stage

After further reorganizing the creep testing data, the creep process can be divided into two stages, that is, the progressive cumulative deformation stage (the axial pressure did not reach the long-term strength) and the sudden failure stage (the axial pressure exceeded the long-term strength).

For the progressive cumulative deformation stage, the trends corresponding to the stable deformation values under each constant axial pressure were compared (Figure 5.8). There was a clear gap between the deformation of

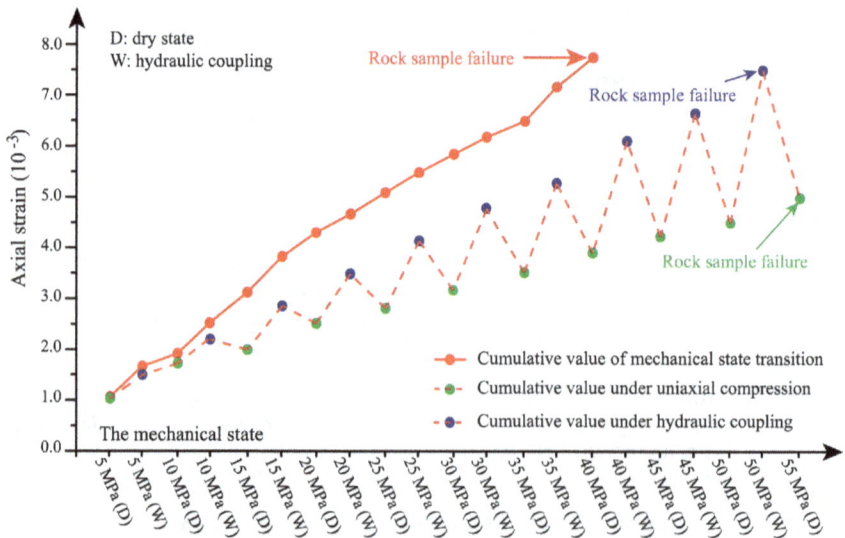

Figure 5.8 The progressive cumulative deformation stage of the creep testing data.

the uniaxial compression state and the hydraulic coupling state under the same axial pressure. As the axial pressure increased, this difference gradually increased. This phenomenon was consistent with the experimental results in Section 5.2. Due to this difference in different mechanical states, the evolution speed and the cumulative deformation of the mechanical transition were significantly greater than those of a single mechanical state.

In addition, since the creep test can provide sufficient time for the stress adjustment of each mechanical state transition, the progressive cumulative deformation can be relatively effective. And after the mechanical state transition, the macroscopic strength of the rock mass was significantly reduced.

5.3.3 Analysis of the sudden failure stage

To quantitatively analyze the sudden failure stage of the sample, based on the experimental data, a constitutive creep model considering the damage evolution was constructed.

The constitutive model used in this section is composed of an elastoplastic damage body and a *Kelvin* body, which can characterize the entire nonlinear accelerated creep evolution of the rock mass (Figure 5.9; Cao et al., 2013). The total strain of rock ε is composed of the strain of elastoplastic damage body ε_{ep} and the strain of *Kelvin* body strain ε_{ve}:

$$\varepsilon = \varepsilon_{ep} + \varepsilon_{ve} \tag{5.1}$$

And the strain of the *Kelvin* body ε_{ep} can be determined by the following formula (Cai et al., 2006):

$$\varepsilon_{ve} = \frac{\sigma_0}{E_1}\left[1 - \exp\left(-\frac{E_1}{\eta}t\right)\right] \tag{5.2}$$

where σ_0 is the axial pressure; E_1 is the viscoelastic modulus, η is the viscoelastic coefficient; and t is the creep time.

For the elastoplastic damage micro-elements, it is composed of damaged and undamaged materials. The cross-sectional areas of the damaged and undamaged materials are set to A_1 and A_2, respectively; and the cross-sectional area of the micro-element of the elastoplastic damage body is A. Therefore, the following formula can be obtained.

$$A = A_1 + A_2 \tag{5.3}$$

Assuming that the nominal stress (i.e., the average stress) of the elastoplastic damage body is σ_0; the damaged part of the material cannot bear the load, and the load is entirely borne by the undamaged part, and the stress is σ_0'.

Therefore, according to the force balance and geometric relationship of the micro-element, the following formula can be obtained:

$$\sigma_0 A = \sigma_0' A_2 \tag{5.4}$$

Define the damage variable D to satisfy the following formula:

$$D = 1 - \frac{A_2}{A} \tag{5.5}$$

Therefore, Equation (5.4) can be changed to:

$$\sigma_0 = \sigma_0' \ (1-D) \tag{5.6}$$

Since σ_0' is the stress suffered by the undamaged material, assuming that its stress–strain relationship follows the linear-elastic Hooke's law, the total strain of the micro-element is ε_{ep}, and the strain of undamaged material is ε_{ep}'. The following formula can be obtained:

$$\sigma_0' = E_0 \varepsilon_{ep}' \tag{5.7}$$

where E_0 is the deformation modulus of elastoplastic damage body. Since the damaged and undamaged parts are closely mixed together, the deformation of the two parts must be coordinated. Therefore, according to the principle of deformation coordination, ε_{ep}' is equal to ε_{ep}. And Equation (5.7) can be changed to:

$$\sigma_0' = E_0 \varepsilon_{ep} \tag{5.8}$$

After substituting Equation (5.8) into Equation (5.6), the following equation can be obtained:

$$\sigma_0 = E_0 \varepsilon_{ep} (1-D) \tag{5.9}$$

Equation (5.9) is the damage constitutive model of rock based on the strain equivalence hypobook proposed by Lemaitre (1984). Using the research results proposed by Kachano (1992), the creep damage rate of rock can be obtained:

$$\frac{dD}{dt} = C \left(\frac{\sigma_0}{1-D} \right)^V \tag{5.10}$$

where C and V are rock material parameters. And the relationship between creep damage variable and creep time satisfies the following formula:

$$D = 1 - \left(1 - \frac{t}{t_{max}}\right)^{\frac{1}{V+1}} \tag{5.11}$$

where t_{max} is the moment when the sample fails during the creep test. According to the existing research results (Cai et al., 2006; Yuan et al., 2006), it is assumed that when the stress is greater than the rock yield limit σ_s, the creep damage of the rock will obviously accumulate. Therefore, the formula for the damage variable is changed to:

$$D = \begin{cases} 0 & \sigma_0 < \sigma_s \\ 1 - \left[1 - \dfrac{t}{t_{max}}\right]^a & \sigma_0 > \sigma_s \end{cases} \tag{5.12}$$

where $a = \dfrac{1}{(V+1)}$. After substituting Equation (5.12) into Equation (5.9), the constitutive model of elastoplastic damage body can be obtained:

$$\varepsilon_{ep} = \begin{cases} \dfrac{\sigma_0}{E_0} & \sigma_0 < \sigma_s \\ \dfrac{\sigma_0}{E_0}\left(1 - \dfrac{t}{t_{max}}\right)^{-a} & \sigma_0 > \sigma_s \end{cases} \tag{5.13}$$

For the sudden failure stage studied in this section, the damage evolution constitutive formulas are rearranged as follows:

$$\varepsilon(t) = \varepsilon_0 (1-D)^{-1} + \frac{\sigma_0}{E_1}\left[1 - \exp\left(-\frac{E_1 t}{\eta}\right)\right], \tag{5.14}$$

$$D = 1 - \left(1 - \frac{t}{t_{max}}\right)^a, \tag{5.15}$$

where $\varepsilon(t)$ is the corresponding strain at time t; ε_0 is the instantaneous elastic strain when $t = 0$; D is the damage variable characterizing the degree of damage evolution of the sample; a is the parameter of the damage variable; t_{max} is the moment when the sample fails: σ_0 is the axial pressure; and E_1 and η are the viscoelastic parameters.

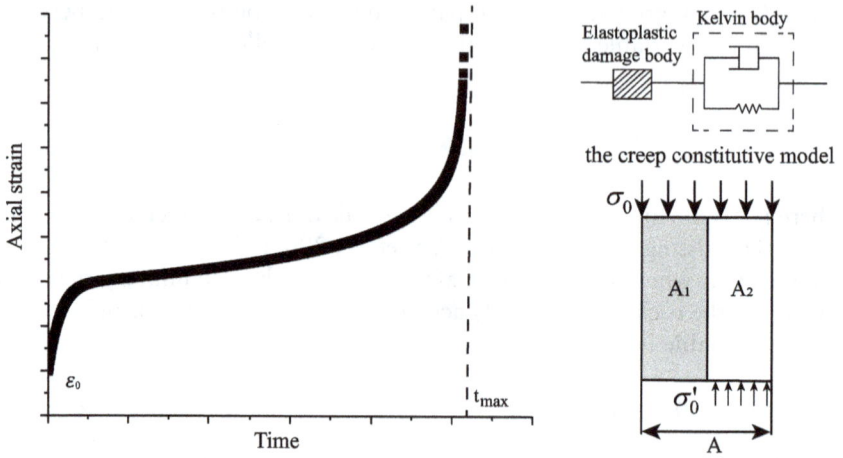

Figure 5.9 The constitutive model of the damage evolution (schematic diagram).

For the determination of these parameters, ε_0, t_{max} and σ_0 can be obtained by the test curve directly, the other parameters can be obtained according to the method mentioned in Sections 5.3.3.1 and 5.3.3.2.

5.3.3.1 Determination of parameter a

According to the creep characteristics of the rock, when the rock entered the accelerated creep stage, the viscoelastic deformation was basically stabilized, while the permanent deformation caused by the damage of the rock continued to increase. Therefore, the constant value of viscoelastic deformation can be obtained by taking the limit of the *Kelvin* body constitutive model. Two points are selected on the test curve in the accelerated creep stage: point (t_i, ε_i) and point (t_j, ε_j). And the parameter a can be solved by the following formula:

$$\varepsilon_i - \varepsilon_j = \varepsilon_0 \left[\left(1 - \frac{t_i}{t_{max}}\right)^{-a} - \left(1 - \frac{t_j}{t_{max}}\right)^{-a} \right] \tag{5.16}$$

5.3.3.2 Determination of parameter E_1 and η

The n test points are selected in the accelerated creep section (t_i, ε_i) $(i = 1, 2, \ldots, n)$, and the viscoelastic modulus E_{1i} of each test point can be obtained by the following formula:

$$E_{1i} = \frac{\sigma_0}{\varepsilon_i - \varepsilon_0 \left(1 - \frac{t_i}{t_{max}}\right)^{-a}} \tag{5.17}$$

And the viscoelastic modulus E_1 is the average of E_{1i}:

$$E_1 = \frac{1}{n}\sum_{i=1}^{n} E_{1i} \qquad (5.18)$$

The n test points on the curve are selected before the accelerated creep (t_i, ε_i) $(i=1, 2, \ldots, n)$, and the viscoelastic coefficient η_i of each test point can be obtained by the following formulas:

$$\eta_i = \frac{E_1 t_i}{\ln \dfrac{\sigma_0}{\sigma_0 - AE_1}} \qquad (5.19)$$

$$A = \varepsilon_i - \varepsilon_0 \left(1 - \frac{t_i}{t_{max}}\right)^{-a} \qquad (5.20)$$

And the viscoelastic coefficient η is the average of η_i:

$$\eta = \frac{1}{n}\sum_{i=1}^{n} \eta_i \qquad (5.21)$$

Based on the test results, the relevant parameters of the constitutive damage model can be obtained (shown in Table 5.2).

According to the fitted evolution curve and the experimental data (Figure 5.10), it can be found that the damage evolution model was very effective. Therefore, the relevant creep parameters can provide an important reference for determining the evolution characteristics of the sample.

In the sudden failure stage, the axial pressure exceeded the long-term strength of the sample, and the creep tests all experienced four stages: instantaneous elastic deformation, decelerated creep, stable creep, and accelerated creep.

Comparing the failure moment during the creep tests (t_{max}), it can be found that the evolution speed under the mechanical state transition was the fastest,

Table 5.2 The parameters of the constitutive damage model

Mechanical environment	σ_0 (MPa)	ε_0 (10^{-3})	t_{max} (h)	E_1 (GPa)	η (GPa · h)	a
Uniaxial creep	55	4.50	21.26	179.192	44.010	0.02355
Creep under hydraulic coupling	50	6.70	18.67	84.783	21.937	0.02763
Creep experiment with mechanical state transition	40	7.18	10.29	74.595	20.775	0.02936

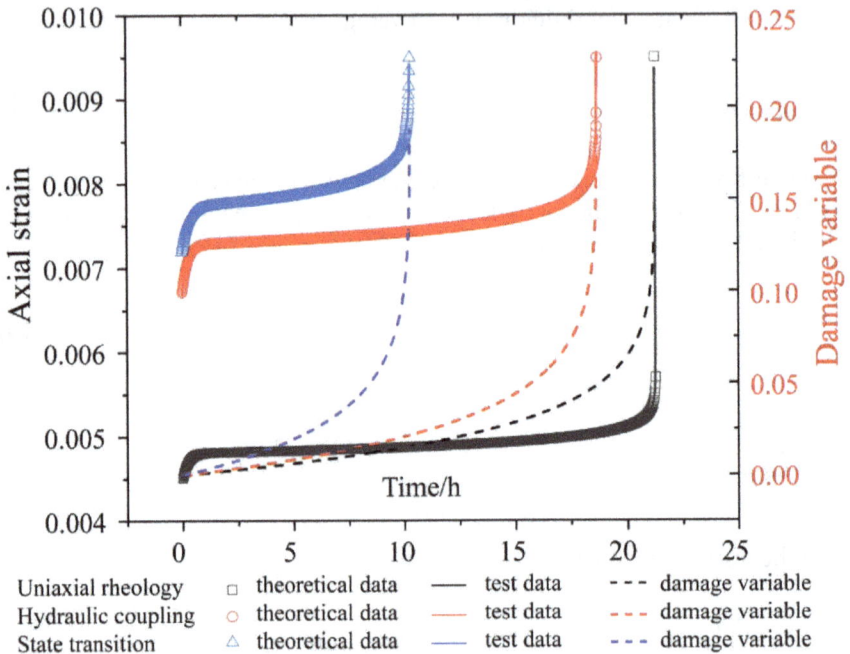

Figure 5.10 Damage evolution in the sudden failure stage.

and the hydraulic coupling state and the uniaxial compression state followed in order. According to the evolution trend of the damage variable (D), the evolution processes under the three mechanical states were both characterized by nonlinear acceleration. Among them, the damage accumulation rate under mechanical state transition was the fastest.

The test data under the mechanical state transition was taken as a case study to conduct the sensitivity analysis of three parameters, involving the viscosity coefficient η, viscoelastic module E_1 and the parameter of the damage variable a. After keeping other parameters unchanged, these three parameters were adjusted separately. The parameter settings are shown in Table 5.3, and the comparison of theoretical curves is shown in Figure 5.11.

Based on the comparison results of the constitutive damage model parameters (Figure 5.11), the viscoelastic parameter E_1 corresponded to the evolution characteristics involving instantaneous elastic deformation, decelerated creep, and stable creep. Specifically, the value of η under the uniaxial creep was the smallest, which indicated that the increase in the magnitude of the instantaneous deformation and the cumulative rate of displacement during the stable creep stage were both largest. In addition, the parameter of the damage variable a and the viscoelastic parameter η corresponded to the evolution

Table 5.3 The parameter settings for sensitivity analysis

Parameters	Initial state	Adjust the viscoelastic module		Adjust the viscosity coefficient		Adjust the parameter of the damage variable	
		Multiply by 2.5	Multiply by 5	Multiply by 2.5	Multiply by 5	Multiply by 2.5	Multiply by 5
E_1 (GPa)	74.595	186.487	372.975	74.595		74.595	
η (GPa · h)	20.775	20.775		51.9375	103.875	20.775	
a	0.02936	0.02936		0.02936		0.0734	0.1468

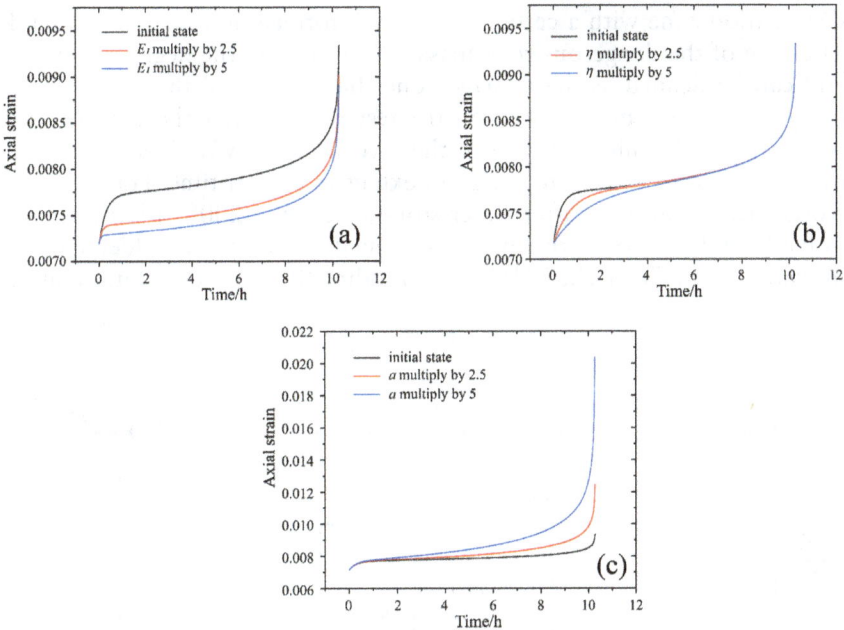

Figure 5.11 The comparison of theoretical curves: (a) adjustment of the viscoe-lastic module; (b) adjustment of the viscosity coefficient; (c) adjustment of the parameter of the damage variable.

characteristics of accelerated creep. The value of η under the mechanical state transition was the largest, which indicated that the evolution rate of accelerated creep was the largest. The value of E_1 under the mechanical state transition was the smallest, which indicated that the increasing magnitude of the speed from the stable creep to the accelerated creep was the largest. Similarly, the related trends in the creep under hydraulic coupling and uniaxial creep decreased in order.

5.4 Evolution process of the dangerous rock mass on the reservoir bank

Based on the test results and on-site monitoring data, the results of previous studies of the JDRM (Huang et al., 2016a; Wang et al., 2020a, 2020b, 2020c), and the results of related studies of similar dangerous rock masses (Feng et al., 2014a, 2014b), especially the dynamic collapse process of Zengziyan (He et al., 2019a, 2019b), the evolution of the dangerous rock mass on the reservoir bank was proposed (Figure 5.12).

5.4.1 Formation of the damage zone (stage A)

Due to the periodic rise and fall of the water level in the reservoir area, a degradation zone with a certain height was formed on the reservoir bank. The base of the dangerous rock mass was located in the degradation zone and can be defined as the damage zone that controlled the overall instability of the rock mass. Based on the test results, after the damage zone was formed, the initial change in the mechanical environment will cause the rock mass to deform to a certain extent in a short time. For the Three Gorges Reservoir area, after water storage began in 2008, a large number of bank landslides occurred in a short time induced by the sudden increase in water level (Yin et al., 2016), which indirectly proved the rationality of

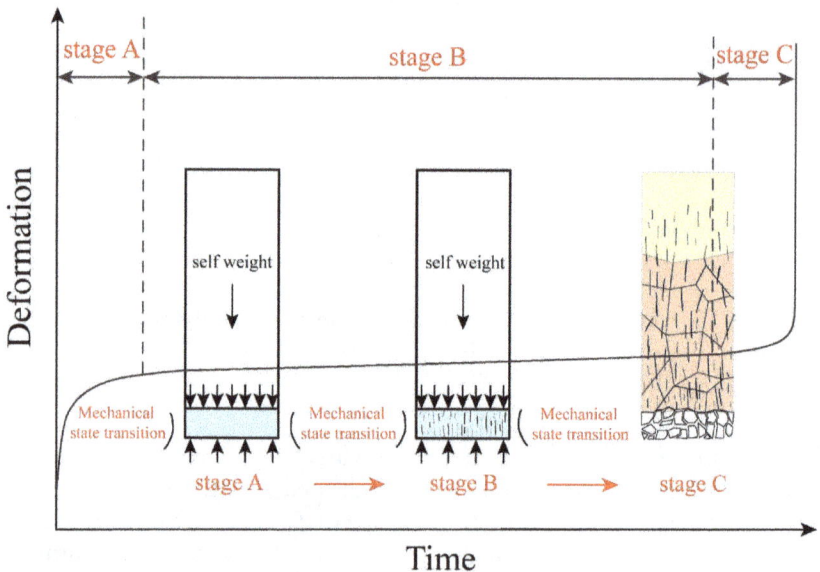

Figure 5.12 The evolution process of the dangerous rock mass on the reservoir bank.

the classification at this stage. After the initial stress adjustment due to the change in the mechanical environment ends, the deformation rate decreased gradually.

5.4.2 Progressive cumulative deformation (stage B)

After the damage zone was formed, it was under the action of the self-weight load of the upper part of the rock mass for a long time. In addition, it periodically changed from the dry state to the hydraulic coupling state. The experimental results of this chapter indicated that two factors caused the dangerous rock mass to enter a critical state during this process. As the deformation increased, the degree of the bias became increasingly serious, resulting in increasing pressure on the base of the rock mass, that is, the axial pressure increased with each hydrological cycle. The change of the mechanical state caused the peak strength of the base of the rock mass to decrease. At this stage, the progressive displacement continued to accumulate at a steady speed.

5.4.3 Nonlinear accelerated failure (stage C)

When the axial pressure was greater than the long-term strength of the base of the rock mass, the base of the rock mass will undergo nonlinear accelerated failure. Compared with the evolution failure under a single mechanical state, since the progressive cumulative deformation completed by the previous periodic mechanical transformation was higher, the evolution rate in stage C was faster (Figure 5.10). The failure process was characterized by nonlinear acceleration (Figure 5.12), which eventually led to the instability of the entire dangerous rock mass in a short time.

5.5 Summary of this chapter

In this chapter, a series of experiments were used to study the influence of reservoir water level on the evolution process of the submerged dangerous rock mass. Conclusions of this chapter can be summarized as follows:

(1) According to the conventional dry–wet cycle test and numerical simulation, it can be found that the dry–wet cycle had little effect on the dense limestone used in this book. However, when the mechanical state was cycled, the difference in rock plasticity and peak strength under different mechanical conditions will become critical to accelerating the deterioration of rock mass.

(2) When considering both axial pressure (self-weight of upper rock mass) and mechanical state transition (the fluctuation of the reservoir water), the peak compressive strength of the dense limestone was lower than that under a single factor of dry–wet cycles. Besides, it can be found that if the

weight of the overlying rock mass were higher, the influence of mechanical transition on the base rock mass would be more significant.

(3) Compared with the creep experiments in a single state, the transition of the mechanical state considering the time-dependent effect continued to promote the degradation of the rock mass, greatly reduced the peak strength of the rock mass, and accelerated the failure of the rock mass. Furthermore, since the creep test can provide sufficient time for the stress adjustment of each mechanical state transition, the progressive deformation can be relatively effective. And then, the macroscopic strength of the rock mass would be significantly reduced.

(4) Based on the field investigation and test results, the entire evolution of the dangerous rock mass on the reservoir bank can be divided into three stages: formation of the damage zone, progressive cumulative deformation, and nonlinear accelerated failure. Specifically, the periodic rise and fall of the reservoir water level led to the formation of a damage zone. Then, the periodic transition of the mechanical state further promoted progressive deformation and reduced the peak intensity of the base of the rock mass, which caused the dangerous rock mass to enter a critical state faster. When the axial pressure exceeded the long-term strength, the evolution of this period nonlinearly accelerated.

Chapter 6

The evolution analysis and the treatment of JDRM

Based on the support of previous test data and the use of the statistical damage constitutive model, this chapter proposed a new analysis method to study the damage evolution process of the dangerous rock masses on the reservoir bank. Applying this method allowed us to quantify the damage and the stability of the submerged dangerous rock mass over time. Afterward, the JDRM was taken as a case study to carry out a detailed analysis by this method. Furthermore, according to parameters extracted from the evolution analysis, the dynamic failure process of the JDRM was obtained. Finally, the mechanical characteristics and the stability trend of the JDRM after the treatment were studied.

6.1 Analysis method of the dangerous rock mass on the reservoir bank

To improve the applicability of the analysis methods, the evolution process of the dangerous rock mass was further simplified (Figure 6.1a). Specifically, the mechanical state transition was defined as the critical factor that promoted the deterioration of the rock mass on the reservoir bank. Meanwhile, the damage variable was used as the key parameter for connecting the different mechanical states. Combining the simplified mechanical state transition with the changing reservoir water level, the evolution of the dangerous rock mass can be quantitatively analyzed by superimposing damaging variables. Notably, in the simplified model mentioned here, the time-dependent effect involved in Chapter 5 was considered by the rise and fall of the reservoir water level.

The analysis method can be divided into four steps: the transition test of the mechanical state, the construction of the statistical damage constitutive model, the superposition of damage variables, and the determination of the factor of safety (FOS). A detailed discussion of each step is shown in the following paragraphs.

DOI: 10.1201/9781003347163-6

6.1.1   The transition test of the mechanical state

After taking the rock samples of the dangerous rock mass in the degradation zone, the rock's full stress–strain curves under dry–wet cycles are required to obtain by indoor test. If test conditions permit, the cycle between the dry state and the hydraulic coupling state under continuous axial pressure would be better. Specifically, the axial pressure is the self-weight of the overlying rock mass, and the confining pressure of hydraulic coupling is the water pressure on the rocky bank when the reservoir water level is the highest (referring to Section 5.2). It should be noted that different rocks are characterized by different sensitivity to water. This difference is caused by the composition of the rock (Cai et al., 2006). The method mentioned here comprehensively considers the difference in strength and plasticity under different states; thus, it can be applied to different rocks.

6.1.2   The construction of the statistical damage constitutive model

According to the simplified mechanical model of the submerged dangerous rock mass (Figure 6.1a), it is assumed that the thickness of the weak base is h; its highest elevation is H_1, and its lowest elevation is H_2, then $h = H_1 - H_2$. The real-time reservoir water level can be defined as a time-dependent function $h(t)$. Therefore, in the typical adjustment process of the reservoir water level, the weak base rock mass underwent four states: the dry state, the hydraulic coupling state, the transition stage of the rising water level, and the transition state of the declining water level (Figure 6.1b).

Based on the mechanical state's transition test and the theory of statistical damage, the damage constitutive model of the rocky bank after the n times hydrological cycles can be obtained. Specifically, the damage constitutive model in the dry state $(h(t) \le H_1)$ is shown as:

$$
\left\{
\begin{array}{l}
\sigma_1^d(n) = E_n^d \varepsilon_1^d(n) \times (1 - D_n^d) \\[2ex]
D^d(n) = 1 - \exp\left[-\left(\dfrac{E_n^d \varepsilon_1^d(n)}{F_0^d(n)} \right)^{m_n^d} \right] \\[3ex]
m_n^d = \dfrac{1}{\ln\left[\dfrac{E_n^d \varepsilon_{1c}^d(n)}{\sigma_{1c}^d(n)} \right]} \\[3ex]
F_0^d(n) = E_n^d \varepsilon_{1c}^d(n) m_n^{d \; \frac{1}{m_n^d}}
\end{array}
\right.
\tag{6.1}
$$

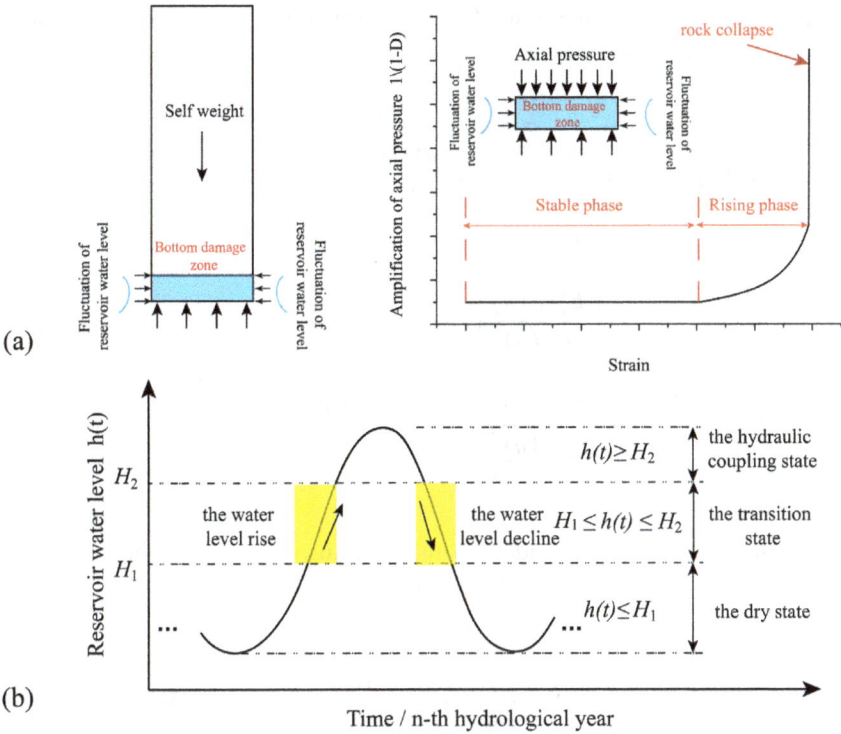

Figure 6.1 (a) the simplified mechanical evolution model (D is the damage variable); (b) the changing mechanical state with reservoir water level.

Source: Yin et al. (2022)

The damage constitutive model under the hydraulic coupling state $(h(t) \geq H_2)$ is expressed as:

$$
\left\{
\begin{array}{l}
\sigma_1^s(n) = E_n^s \varepsilon_1^s(n) \times (1 - D_n^s) \\[2mm]
D^s(n) = 1 - \exp\left[-\left(\dfrac{E_n^s \varepsilon_1^s(n)}{F_0^s(n)} \right)^{m_n^s} \right] \\[4mm]
m_n^s = \dfrac{1}{\ln\left[\dfrac{E_n^s \varepsilon_{1c}^s(n)}{\sigma_{1c}^s(n)} \right]} \\[6mm]
F_0^s(n) = E_n^s \varepsilon_{1c}^s(n) m_n^{s \frac{1}{m_n^s}}
\end{array}
\right.
\tag{6.2}
$$

The damage constitutive model during the transition stage of the water level rise $(H_2 \geq h(t) \geq H_1)$ is stated as:

$$
\begin{cases}
\sigma_1^u(n) = E_n^d \varepsilon_1^d(n) \times (1 - D_n^d) - \dfrac{(h(t) - H_1)}{H_2 - H_1} \\
\qquad \times \left[E_n^d \varepsilon_1^d(n) \times (1 - D_n^d) - E_n^s \varepsilon_1^s(n) \times (1 - D_n^s) \right] \\[2mm]
D^u(n) = 1 - \exp\left[-\left(\dfrac{E_n^d \varepsilon_1^d(n)}{F_0^d(n)} \right)^{m_n^d} \right] + \dfrac{(h(t) - H_1)}{H_2 - H_1} \\[2mm]
\qquad \times \left\{ \exp\left[-\left(\dfrac{E_n^d \varepsilon_1^d(n)}{F_0^d(n)} \right)^{m_n^d} \right] - \exp\left[-\left(\dfrac{E_n^s \varepsilon_1^s(n)}{F_0^s(n)} \right)^{m_n^s} \right] \right\}
\end{cases}
\tag{6.3}
$$

The damage constitutive model during the transition stage of the water level decline $(H_2 \geq h(t) \geq H_1)$ is presented as:

$$
\begin{cases}
\sigma_1^f(n) = E_n^s \varepsilon_1^s(n) \times (1 - D_n^s) + \dfrac{(H_2 - h(t))}{H_2 - H_1} \\
\qquad \times \left[E_n^d \varepsilon_1^d(n) \times (1 - D_n^d) - E_n^s \varepsilon_1^s(n) \times (1 - D_n^s) \right] \\[2mm]
D^f(n) = 1 - \exp\left\{ -\left(\dfrac{E_n^s \varepsilon_1^s(n)}{F_0^s(n)} \right)^{m_n^s} \right\} - \dfrac{(H_2 - h(t))}{H_2 - H_1} \\[2mm]
\qquad \times \left\{ \exp\left[-\left(\dfrac{E_n^d \varepsilon_1^d(n)}{F_0^d(n)} \right)^{m_n^d} \right] - \exp\left[-\left(\dfrac{E_n^s \varepsilon_1^s(n)}{F_0^s(n)} \right)^{m_n^s} \right] \right\}
\end{cases}
\tag{6.4}
$$

In the aforementioned formulas, $\sigma_1(n)$ is axial pressure in the n-th hydrological year; $D(n)$ is damage variable in the n-th hydrological year; $E(n)$ is the elastic modulus in the n-th hydrological year; $\varepsilon_1(n)$ is the real-time strain in the n-th hydrological year; $\varepsilon_{1c}(n)$ is the strain corresponding to peak intensity in the n-th hydrological year; $\sigma_{1c}(n)$ is the peak intensity in the n-th hydrological year; $F_0(n)$ and m_n are the damage constitutive parameters in the n-th hydrological year. The upper right corners represent dry state (d), hydraulic coupling state (s), water level rise (u), and water level fall (f). The parameters involved here can be obtained from the full stress–strain curve under the transition test of mechanical state, and the specific method of model construction will be described in detail later.

6.1.3   The superposition of damage variables

The damage accumulation of rock mass under the action of water level fluctuation can be quantified according to the following formulas:

$$D_E(n-1) = 1 - \frac{\sigma_E(n-1)}{E_{n-1}^s \varepsilon_1^s(n-1)} \tag{6.5}$$

$$\sigma_E(n) = E_n^d \varepsilon_1^d(n) \times \left[1 - D_E(n-1)\right] \tag{6.6}$$

where $\sigma_E(n-1)$ and $\sigma_E(n)$ are the effective stress of the base rock mass after the $(n-1)$ and n hydrological year, respectively. Note that $n \geq 1$, and $\sigma_E(0)$ is the self-weight of the overlying rock mass. $\varepsilon_1^s(n-1)$ is the effective strain corresponding to the effective stress of base rock mass after the $n-1$ hydrological years. E_{n-1}^s is the elastic modulus of the base rock mass after the $n-1$ hydrological years. $D_E(n-1)$ is the effective damage variable after the $n-1$ hydrological years. $\varepsilon_1^d(n)$ is the effective strain corresponding to the effective stress of base rock mass after the n hydrological years. E_n^d is the elastic modulus of the base rock mass after the n hydrological years.

6.1.4   The determination of the FOS

As the number of fluctuations in the water level increases, the effective stress of the base rock mass can be obtained by superimposing the damage variable. And the FOS of the dangerous rock mass can be determined by:

$$F_{s-d} = \frac{\sigma_c(n)}{\sigma_E(n)} \tag{6.7}$$

where F_{s-d} is the FOS that considers the damage evolution under the action of water level fluctuation; $\sigma_c(n)$ is the peak intensity of base rock mass in the n-th hydrological year; $\sigma_E(n)$ is the effective stress of the base rock mass in the n-th hydrological years.

To better understand the application of this method, the JDRM was taken as a case study to carry out a detailed analysis. Besides, sufficient monitoring data of JDRM will be used to verify this method's effectiveness.

6.2   Damage evolution and stability analysis of JDRM

6.2.1   The statistical constitutive damage model under dry–wet cycles

Based on the previous analysis, the JDRM base rock mass was under axial compression and low confining pressure. According to the field investigation (Chapter 2), the transition test of the mechanical state

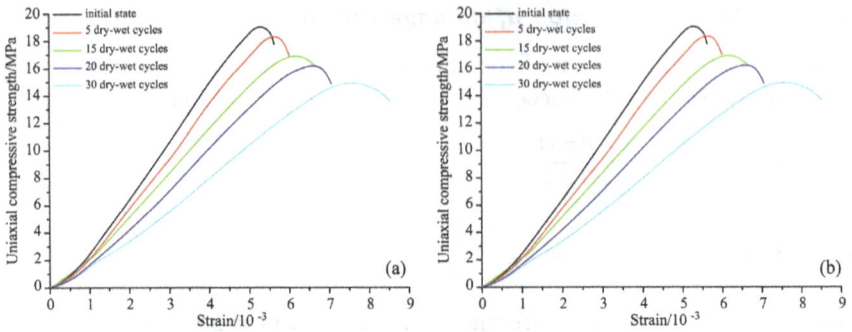

Figure 6.2 Stress–strain curves at different numbers of dry–wet cycles. (a) The dry condition. (b) The saturated condition.

Source: Wang et al. (2020d)

mentioned here was simplified as the dry–wet cycle test (Hudson and Harrison, 2000; Standard for test methods of engineering rock mass, 2013; Ulusay, 2014).

Figure 6.2 shows the stress–strain curves for an environment experiencing dry–wet cycles. There were significant differences in the stress–strain curves for a dry state and a saturated state. When the mechanical environment changed from a dry state to a saturated state, the peak stress intensity was reduced by 30.57–35.63%. As the number of dry–wet cycles increased, the stress reduction between a dry and a saturated state grew, the peak strength decreased, and the degree of rock brittleness gradually decreased (i.e., the strain at the time of the peak strength increased, Hudson and Harrison, 2000; Ulusay, 2014). Due to the mechanical difference between the dry condition and the saturated condition, the dry–wet cycles accelerated the damage imparted to the rock. Therefore, a constitutive damage model should include both the dry condition and the saturated condition.

The constitutive model was defined according to the two parts of the stress–strain curves shown in Figure 6.2. When the strain ε_1 is less than or equal to the peak strain of the compaction section ε_{1cc}, the constitutive model is:

$$\left\{ \begin{array}{l} \sigma_1 = E_{ncc}\varepsilon_1 \exp\left[-\left(\dfrac{E_{ncc}\varepsilon_1}{F_0}\right)^m\right] \\[4mm] D = 1 - \exp\left\{-\left(\dfrac{E_{ncc}\varepsilon_1}{F_0}\right)^m\right\} \end{array} \right. \tag{6.8}$$

When the strain ε_1 is greater than or equal to the peak strain of the compaction section ε_{1cc}, the constitutive model is:

$$
\left\{
\begin{array}{l}
\sigma_1 = E_n\left(\varepsilon_1 - \varepsilon_{1cc}\right)\exp\left[-\left(\dfrac{E_n\left(\varepsilon_1 - \varepsilon_{1cc}\right)}{F_0}\right)^m\right] + \sigma_{1cc} \\[4mm]
D = 1 - \exp\left\{-\left(\dfrac{E_n\left(\varepsilon_1 - \varepsilon_{1cc}\right)}{F_0}\right)^m\right\} + D_c
\end{array}
\right.
\tag{6.9}
$$

In Equations (6.8) and (6.9), σ_1 is the stress, E_{ncc} is the tangent elastic modulus of the compacted section of the curve, σ_{1cc} is the peak strength of the compacted section, D is the damage variable, D_c is the peak damage of the compacted section, m and F_0 are the Weibull distribution parameters of the rock material, and E_n is the elasticity modulus of the rock after n dry–wet cycles, which is obtained by linear fitting. E_n under dry conditions is defined as:

$$
E_{n-d} = -56.772n + 4072.8,
\tag{6.10}
$$

and E_n under saturated conditions is defined as:

$$
E_{n-s} = -55.912n + 3908.8 .
\tag{6.11}
$$

m and F_0 can be obtained using the equations of Liu et al. (2018b):

$$
m = \frac{1}{\ln\left[E_n\varepsilon_{1c}/\sigma_{1c}\right]},
\tag{6.12}
$$

$$
F_0 = E_n\varepsilon_{1c}m^{1/m},
\tag{6.13}
$$

where ε_{1c} is the strain corresponding to the peak intensity and σ_{1c} is the peak strength of the rock mass. The constitutive damage parameters of the rock mass experiencing dry–wet cycles are shown in Table 6.1.

According to the constitutive damage model, the stress–strain curves for rocks experiencing dry–wet cycles can be determined. Compared the theoretically derived curves with the experimental curves, it can be found that the two sets of curves were in strong agreement with one another (Figure 6.3). Therefore, the inclusion of the impact of the dry–wet cycles on the rock mass stress–strain curves was reasonable. Based on the equations mentioned earlier, the constitutive curve and the damage variable for each dry–wet cycle can be calculated, which can determine how the damage progresses over time.

Table 6.1 Constitutive parameters of the rock mass experiencing dry–wet cycles

Dry–wet cycle	σ_{1c}/MPa	ε_{1c} / 10^{-3}	m	F_0	E_{ncc}/ MPa	ε_{1cc} / 10^{-3}	σ_{1cc} / MPa
Initial dry condition	19.07	4.22	17.82	20.23	2.72	1.03	2.80
Dry condition in the 5th dry–wet cycle	18.39	4.46	13.45	20.50	2.53	1.07	2.70
Dry condition in the 15th dry–wet cycle	16.96	5.01	9.10	20.59	2.15	1.16	2.49
Dry condition in the 20th dry–wet cycle	16.22	5.35	7.86	20.43	1.96	1.22	2.38
Dry condition in the 30th dry–wet cycle	14.98	6.34	6.21	20.14	1.58	1.39	2.20
Initial saturated condition	13.24	3.47	5.48	18.49	2.61	0.74	1.94
Saturated condition in the 5th dry–wet cycle	12.25	3.51	5.03	17.58	2.42	0.75	1.80
Saturated condition in the 15th dry–wet cycle	11.10	3.85	4.48	16.54	2.05	0.80	1.63
Saturated condition in the 20th dry–wet cycle	10.44	4.05	4.18	15.92	1.86	0.82	1.53
Saturated condition in the 30th dry–wet cycle	9.67	4.81	3.81	15.23	1.49	0.95	1.42

Source: Wang et al. (2020d)

6.2.2   Mechanical damage analysis of the submerged dangerous rock mass

According to the JDRM model, the weight of the upper part of the rock mass and the periodic change in the reservoir water level led to a decrease in strength and an increase in damage of the base rock mass. The mechanical properties of the base rock mass in a dry state and a saturated state were quite different, and the fluctuation in the mechanical strength between these two states sped up the damage to the base rock mass. Based on the generalized model (Figure 6.1), the actual rock conditions were replicated in an indoor test. The average height of

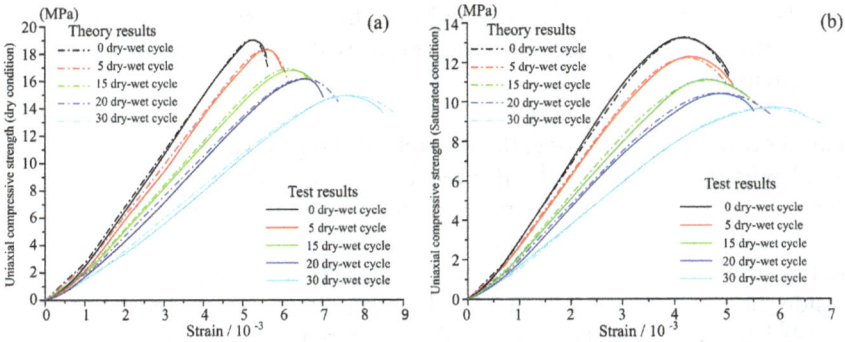

Figure 6.3 Comparison of the experimental and theoretical stress–strain curves for the samples in (a) a dry state and (b) a saturated state.

Source: Wang et al. (2020d)

the upper part of the rock mass was 120 m, the average unit weight was 27.0 kN/m³, and the weight of the upper part of the rock mass was 3.24 MPa, which was close to the peak value observed during monitoring in 2014. Therefore, 3.24 MPa was used as the initial axial pressure for the model. The samples used in the dry–wet cycle tests were assumed to be representative of the mechanical state of the base rock mass. Before the derivation, the following assumptions were made:

(1) The axial pressure of the base rock mass remained unchanged and was set equal to 3.24 MPa. However, as the damage accumulated, the rock destruction continued to accumulate on a smaller scale, and the effective axial pressure increased (Nejati and Ghazvinian, 2014). This assumption was consistent with the observed pressure changes and crack deformation.

(2) According to the laboratory test data, the axial pressure of 3.24 MPa exceeded the peak strength of the compaction section. In addition, the field investigation and monitoring data indicated that the JDRM had already entered the stage of accelerated deformation. Therefore, after considering the key parameters, such as the elastic modulus, peak strength, and corresponding strain, the stress–strain curve was simplified to an elastoplastic model that ignored the damage of the compacted section but maintained the continuity of the damage variables. The constitutive model was simplified as follows:

$$
\begin{cases}
D = 1 - \exp\left\{ -\left(\dfrac{E_n \varepsilon_1}{F_0} \right)^m \right\} \\[4mm]
\sigma_1 = E_n \varepsilon_1 (1 - D) = E_n \varepsilon_1 \exp\left\{ -\left(\dfrac{E_n \varepsilon_1}{F_0} \right)^m \right\}
\end{cases}
\tag{6.14}
$$

(3) The increase in the damage and the effective axial pressure were irreversible, and cannot decrease, even with the changing mechanical environment.

Based on these assumptions, the theoretical damage values and the effective axial pressure values can be derived by substituting the initial axial pressure into the constitutive damage model for the initial dry–wet cycle. For the initially dry rock mass, the damage variable corresponding to the initial axial pressure of 3.24 MPa was 0.00039; this value defined the damage inside the rock mass (Hudson and Harrison, 2000; Nejati and Ghazvinian, 2014; Ulusay, 2014). For the initially saturated rock, the damage variable corresponding to the initial axial pressure of 3.24 MPa was 0.000636, which represented an increase of 38.68% compared to the dry condition damage value. Because the damage was irreversible, the 0.000636 saturated damage value was used as the initial condition for the first dry–wet cycle. Then this calculation iteratively for multiple dry–wet cycles was repeated. By the ninth dry–wet cycle, as the dry rock state transitioned into the saturated rock state, the effective axial pressure exceeded the saturated compressive strength of the rock mass, and it was predicted that the rock mass will be dangerously unstable (Table 6.2).

Comparing the theoretical pressure values to the observed pressure data (Figure 6.4), it can be found that these two nonlinear trends were highly consistent with one another; this similarity validated the generalized JDRM model. In the presence of an axial load with low confining pressure, the strength reduction and the damage evolution of the rock mass as it experienced successive dry–wet cycles were the key factors that determined the

Table 6.2 Evolution of the damage parameters for the base rock mass

Dry–wet cycle	Effective axial pressure (MPa)	Increase percentage of effective axial pressure (%)	Cumulative value of the damage variable	Increase percentage of the damage variable (%)
0	3.24	/	0.000636	/
1	3.82	17.90	0.001144	79.87
2	4.32	13.09	0.002047	78.93
3	4.88	12.92	0.003602	75.96
4	5.576	14.31	0.007743	114.96
5	6.668	19.58	0.015997	106.60
6	7.949	19.21	0.032893	105.61
7	9.555	20.20	0.070953	115.71
8	11.721	22.67	0.195443	175.45
9	15.615	33.22	Rock collapse	Rock collapse

Source: Wang et al. (2020d)

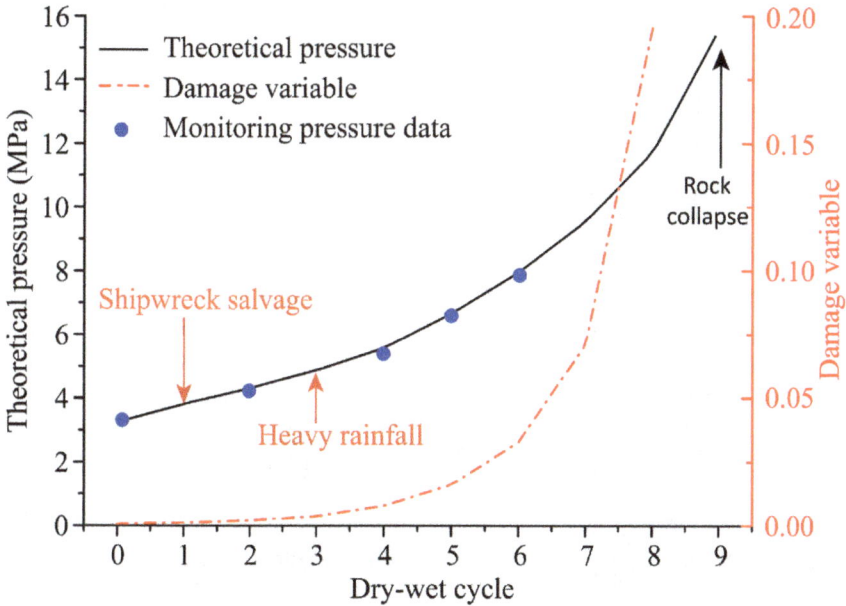

Figure 6.4 Comparison of theoretical and observed pressure values.
Source: Wang et al. (2020d)

overall instability of the JDRM. Unexpected reservoir conditions, such as a shipwreck salvage in June 2015 and a heavy rainfall event in June 2016, could result in an additional decrease in the strength of the rock mass above and beyond that induced by the dry–wet cycles. During these two abnormal periods, the rock mass experienced the strength reduction of two dry–wet cycles in a hydrological year. Although the deterioration of the rock mass involved a stepwise increase in the abnormal environment, the deterioration trend remained unchanged.

As of 2019, the peak pressure value observed at the monitoring station was 7.89 MPa. According to the theoretical deterioration trend, the dangerous rock mass will reach a critical stage after two more dry–wet cycles; reinforcement or other engineering support methods must be implemented as soon as possible.

6.3   Stability analysis of JDRM

To quantify the deterioration trend of the dangerous rock mass, the stability trend of the JDRM as it experienced successive dry–wet cycles were calculated. For these calculations, the conventional static methods (i.e., Geo-hazard Prevention and Management Command of the Three Gorges Reservoirs,

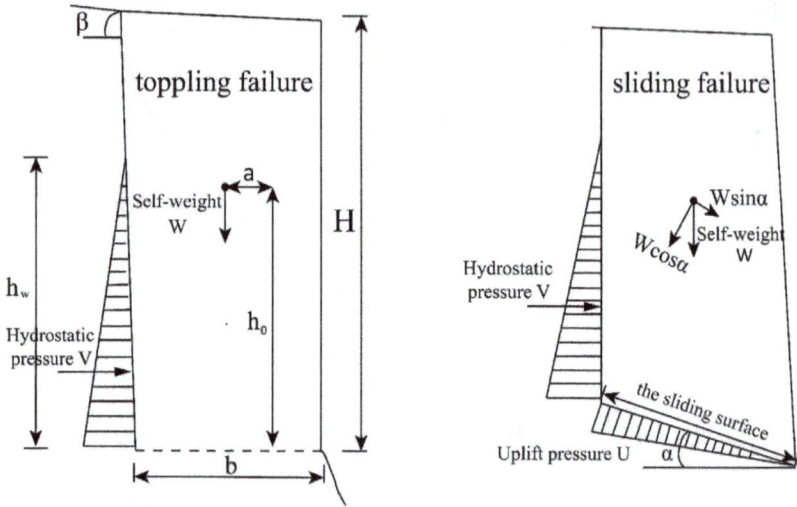

Figure 6.5 Traditional stability calculations for two different dangerous rock mass failure modes.

Source: Wang et al. (2020d)

2004) were employed that define two basic modes of rock mass collapse: toppling failure and sliding failure (Figure 6.5). The calculation used for toppling failure is:

$$F_{s-t} = \frac{\frac{1}{2} f_{lk} \times b^2 + W \times a}{V\left(\frac{h_w}{3\sin\beta} + b\cos\beta\right)},$$

(6.15)

and the calculation for sliding failure is:

$$F_{s-s} = \frac{(W\cos\alpha - V\sin\alpha - U)\tan\varphi + cl}{W\sin\alpha + V\sin\alpha}.$$

(6.16)

In Equations (6.14) and (6.15), F_{s-t} is the FOS of the toppling failure, F_{s-s} is the FOS of the sliding failure, c is the cohesion of the sliding surface (kPa), φ is the internal friction angle of the sliding surface (°), l is the length of the sliding surface (m), W is the weight of the dangerous rock mass (kN/m), α is the inclination of the sliding surface (°), h_w is the height of the water filling the fracture (m), V is the water pressure at the trailing edge of the fissure (kN/m), U is the uplift pressure on the slip surface (kN/m), a is the horizontal distance from the center of gravity of the dangerous rock mass to the overturning point (m), b is the horizontal distance from the lower end of the trailing edge

Table 6.3 The constant mechanical parameters used in the JDRM stability model

b (m)	h_w (m)	β (°)	W (kN/m)	a (m)	V (kN/m)	α (°)	l (m)	U (kN/m)
28.5	65	80	150117.34	20.43	21125	7	28.61	9298.25

Source: Wang et al. (2020d)

Table 6.4 The results of the traditional stability model calculations

Dry–wet cycle	f_{lk} (kPa)	c (kPa)	Tan φ	F_{s-t}	F_{s-s}
Initial dry condition	220	472	0.64	5.99	4.85
Initial saturated condition	152.74	377.6	0.512	5.94	3.88
Dry condition in the 5th dry–wet cycle	198	426	0.63	5.97	4.72
Saturated condition in the 5th dry–wet cycle	131.89	340.8	0.504	5.92	3.78
Dry condition in the 15th dry–wet cycle	178	383	0.62	5.95	4.60
Saturated condition in the 15th dry–wet cycle	116.49	306.4	0.496	5.91	3.68
Dry condition in the 20th dry–wet cycle	166	358	0.58	5.95	4.30
Saturated condition in the 20th dry–wet cycle	106.84	286.4	0.464	5.90	3.44
Dry condition in the 30th dry–wet cycle	158	339	0.54	5.94	4.01
Saturated condition in the 30th dry–wet cycle	101.99	271.2	0.432	5.89	3.21

Source: Wang et al. (2020d)

to the overturning point (m), f_{lk} is the tensile strength of the dangerous rock mass (kPa), and β is the inclination angle of the trailing edge crack (°).

The parameters used in and the results from the traditional stability analysis are shown in Tables 6.3 and 6.4, respectively. The values of the parameters were reduced according to the published standard (Geo-hazard Prevention and Management Command of the Three Gorges Reservoirs, 2004): the compressive strength of the rock mass was multiplied by the influence coefficient of the crack (0.33), the tensile strength and cohesion of the rock mass were multiplied by the reduction factor (0.20), the internal friction angle was multiplied by the reduction factor (0.80), and the rock deformation parameter was multiplied by the reduction factor (0.70).

After 30 dry–wet cycles, the FOS values for the toppling failure and the sliding failure decreased from the initial stability values by 1.67% and 36.04%, respectively (Table 6.4). However, these static calculations showed that the dangerous rock mass will be stable after thirty dry–wet cycles, which was not consistent with the current pressure monitoring data and the crack development trend.

Based on the aforementioned analysis, the deterioration of the base rock mass was the key to the instability of the JDRM. In addition, the mechanical model of the base rock mass can be generalized to a rock mass sample under constant axial pressure with progressive deterioration driven by multiple dry–wet cycles. Therefore, the FOS of the JDRM can be quantitatively analyzed based on the damage degradation caused by the dry–wet cycles. The stability calculation based on the damage evolution can be obtained by Equation (6.7). The FOS results, based on the damage evolution of the JDRM, are shown in Table 6.5.

The FOS fluctuates cyclically with the transition from a dry state to a saturated state. To ensure that the FOS changed continuously with the water level in the reservoir area, the saturated (water level higher than 160 m) and dry (water level lower than 150 m) states were defined as instances when the water level was higher or lower than the marlstone, respectively. The FOS of the JDRM was set to change linearly at other water levels. The final FOS calculations are shown in Figure 6.8. In order to highlight the variation in the

Table 6.5 FOS results for the damage evolution of the JDRM

State of rock mass	Effective compressive strength of the rock mass (MPa)	Compressive strength of the base rock mass (MPa)	F_{s-d}
Initial dry condition	3.240	19.053	5.880
Initial saturated condition		12.989	4.009
1st dry condition	3.820	18.915	4.952
1st saturated condition		12.871	3.369
2nd dry condition	4.320	18.777	4.347
2nd saturated condition		12.753	2.952
3rd dry condition	4.878	18.639	3.821
3rd saturated condition		12.635	2.590
4th dry condition	5.576	18.502	3.318
4th saturated condition		12.518	2.245
5th dry condition	6.668	18.364	2.754
5th saturated condition		12.400	1.860
6th dry condition	7.949	18.226	2.293
6th saturated condition		12.282	1.545
7th dry condition	9.555	18.088	1.893
7th saturated condition		12.164	1.273
8th dry condition	11.721	17.951	1.531
8th saturated condition		12.047	1.028
9th dry condition	15.615	17.813	1.141
9th saturated condition		11.929	0.764

Source: Wang et al. (2020d)

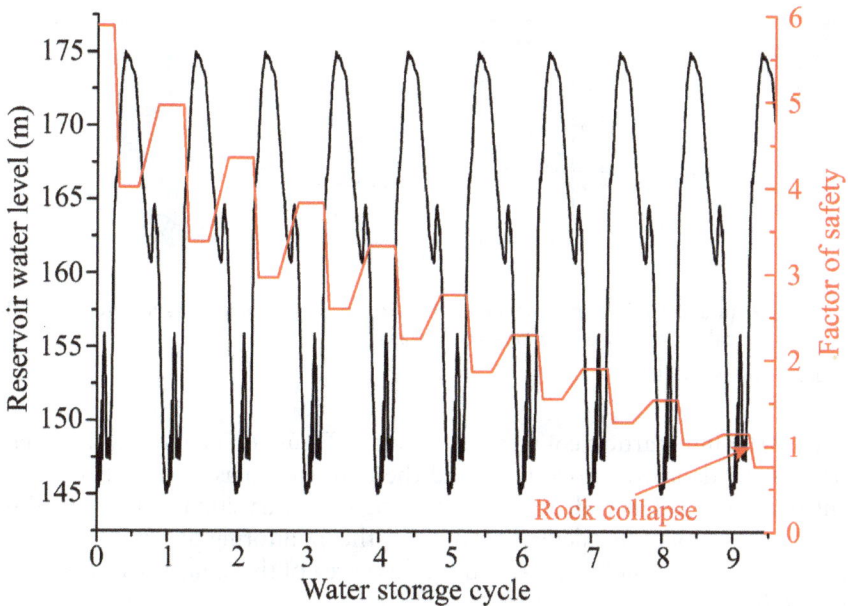

Figure 6.6 FOS results superimposed on the water level in the reservoir area.

Source: Wang et al. (2020d)

FOS with water level, the minor water level fluctuations were ignored that occurred over the years during the calculations.

Figure 6.6 showed that the FOS values decreased with an increasing number of water storage cycles and that the JDRM will fail during the transition from the dry state to the saturated state for the ninth water storage cycle. Due to differences in the strength of the base rock mass between the dry state and the saturated state, the FOS fluctuated considerably. As the number of dry–wet cycles increased, the amplitude of the fluctuation gradually decreased, and the impact of the difference between the dry state and the saturated state on the FOS fluctuation became smaller. Compared to the conventional failure analysis, this quantitative method yielded results that were closer to the observed values.

6.4 Analysis of dynamic collapse of JDRM

According to the critical state determined by the previous method, the relevant parameters were extracted to analyze the dynamic collapse of the JDRM.

6.4.1 Establishment of the numerical model

Based on the typical section of the JDRM, a two-dimensional numerical model of the JDRM was established. The model contained 36989 particles

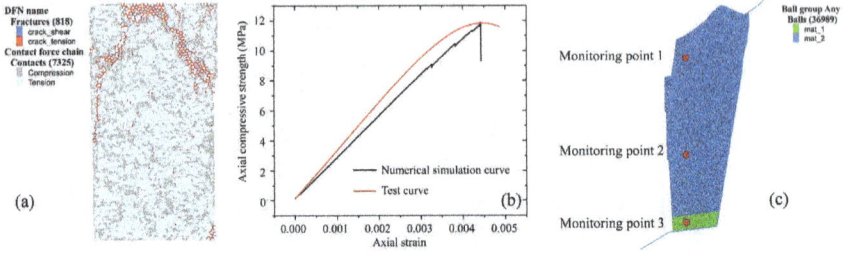

Figure 6.7 (a and b) The parameter calibration of base rock mass and (c) numerical model of dangerous rock mass.

Source: Yin et al. (2022)

and 89566 interparticle contacts, as shown in Figure 6.7(a and b). The upper part of the dangerous rock mass and the base rock mass were set to different parameters. The bedrock at the trailing edge was simplified as a wall to reduce the amount of calculation. Besides, the monitoring points were placed in the upper part, middle part, and the base area of the dangerous rock mass, respectively.

6.4.2   Determination of parameters

According to the previous theoretical analysis, the JDRM will enter a critical state after nine dry–wet cycles. At this time, the effective stress of the overlying rock mass was 15.615 MPa. During the numerical simulation, the damage of the base rock mass was generalized to the increase of the self-weight of the upper rock mass. The deterioration of the base rock mass was completed by adjusting the microscopic parameters. Therefore, the microscopic parameters of the base rock mass were calibrated according to the macroscopic parameters of the saturated rock mass after the nine dry–wet cycles, as shown in Figure 6.7c. Besides, the microscopic parameters of the upper rock mass were the same as the uniaxial compression state in Chapter 4. The related micro-parameters are shown in Table 6.6.

Table 6.6 Microscopic parameters in dynamic collapse analysis

Rock mass	Effective modulus (GPa)	Bond effective modulus (GPa)	Normal-to-shear stiffness ratio	Bond normal-to-shear stiffness ratio	Tensile strength (MPa)	Cohesion (MPa)
Base area	1.44	1.74	1.63	1.0	4.85	4.85
Upper part	6.518	6.615	1.2	1.2	28.49	23.74

Source: Yin et al. (2022)

6.4.3   Dynamic analysis of collapse

The rock mass collapse process under different calculation steps was obtained through numerical calculation, as shown in Figure 6.8.

By analyzing the dynamic collapse process of the dangerous rock mass, the following conclusions can be obtained. Before the collapse, cracks were preferentially concentrated on the water-side of the base area (Figure 6.8a). This phenomenon was caused by a certain bias degree of self-weight pressure on the waterfront side, leading to preferential failure. After that, the fractures in the base area gradually increased, and the fractures extended upward from the base area (Figure 6.8b–d). This failure mode was consistent with the

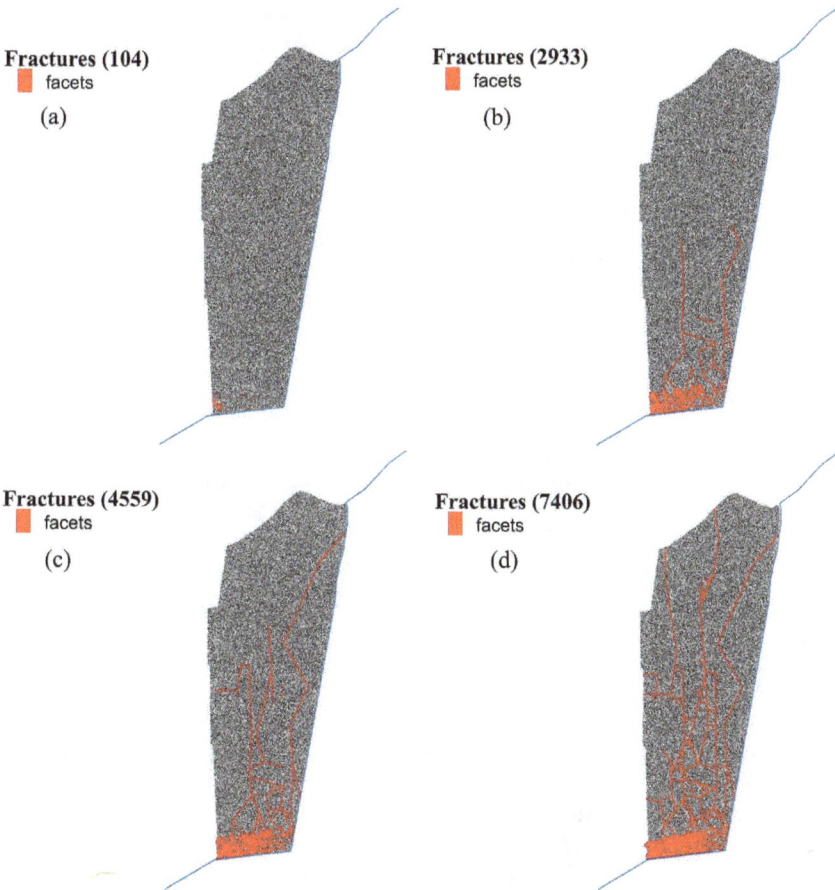

Figure 6.8 Dynamic process of collapse: (a) Step 2000; (b) Step 6000; (c) Step 10000; (d) Step 20000; (e) Step 500000; (f) Step 2000000.

Source: Yin et al. (2022)

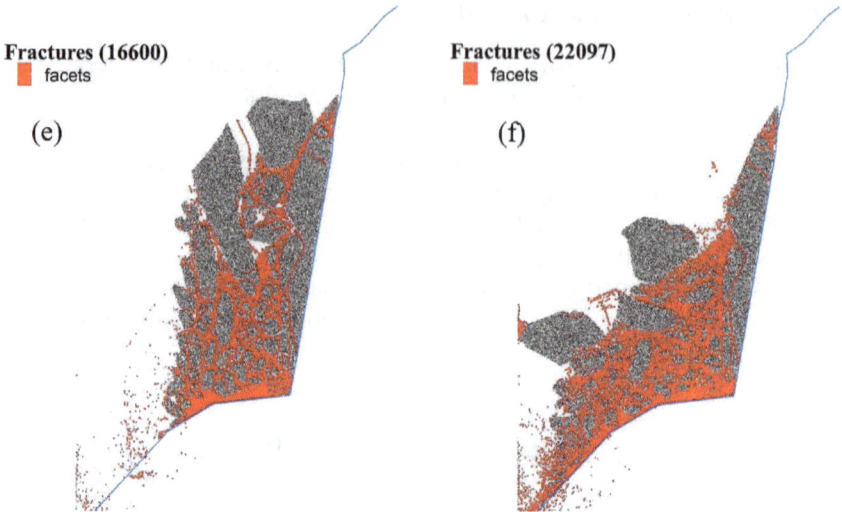

Fractures (16600)
- facets

(e)

Fractures (22097)
- facets

(f)

Figure 6.8 (Continued)

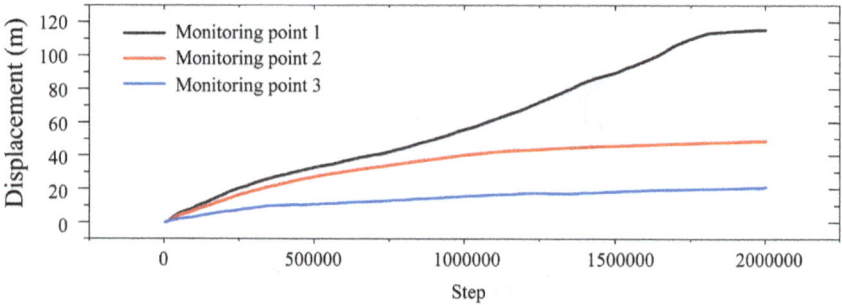

Figure 6.9 Displacement curve of monitoring points.

Source: Yin et al. (2022)

theoretical model obtained earlier. After the cracks penetrated the dangerous rock mass, the upper part disintegrated and the rock mass collapsed. During the collapse, the waterfront displacement of the JDRM was large, while the displacement on the trailing edge was small. The displacement curves of the monitoring points are shown in Figure 6.9.

By analyzing the displacement curves at different positions, it can be found that the displacement of the upper rock mass was the largest, and the displacements of the middle rock mass and the base area decreased sequentially. Due to the limited space for movement, the rock mass in the base area entered the stable state first, and its maximum displacement was 21.58 m. After that, the middle rock mass came to a stable state with a maximum displacement of

48.9 m. And the upper rock mass entered a stable state with a maximum displacement of 115.41 m. Moreover, the displacement trends of the monitoring points at different positions were similar; that was, the increasing speed of the displacement gradually decreased and eventually dropped to zero. This failure process was identical to the dynamic characteristics of the Zengziyan collapse (He et al., 2019a, 2019b).

6.5   Analysis of the treatment of JDRM

6.5.1   Preventative methods for the JDRM

According to the deformation characteristics and failure modes of the JDRM, the deterioration of the base rock mass and the deformation caused by uneven pressure were the key factors controlling the instability of the entire dangerous rock mass. Therefore, the disaster prevention plan for the JDRM was designed in two parts: the reinforcement of the weak base of the rock mass and the anchoring of the middle and upper parts of the dangerous rock mass. The reinforcement of the base of the rock mass was performed to improve the strength of the base rock mass. And the anchoring of the middle and upper parts of the dangerous rock mass was performed to control the deformation of the dangerous rock mass. The layout of these control measures is shown in Figure 6.10a, and the specific protective measures were as follows:

(1) The reinforcement of the base of the rock mass. The adit was supported by and filled with C30 concrete. The base of the rock mass was reinforced with fully grouted bolts. Where the spacing between the bolts was 1.75 m and 2.25 m, the diameter of the bolt was 150 mm, and the length of the anchored section was 6 m. The surface of the submerged base of the rock mass was reinforced with lattice beams.
(2) The anchorage of the middle and upper parts of the dangerous rock mass. Six rows of anchor cables were arranged within the upper part of the dangerous rock mass, the horizontal angle was 15°, the horizontal spacing was 6.00 m, and the vertical spacing was 6.00 m. The total length of the anchored section was 17.00 m.

As of January 2019, the preventative methods for the JDRM were being implemented in accordance with the established plan (Figure 6.10b).

6.5.2   Influence of the preventative methods

To study the influence of the preventative methods on the JDRM, the displacement field and the stress field of the dangerous rock mass were compared before and after the application of the preventative methods using a numerical simulation. It should be noted that in the numerical calculation, the mechanical

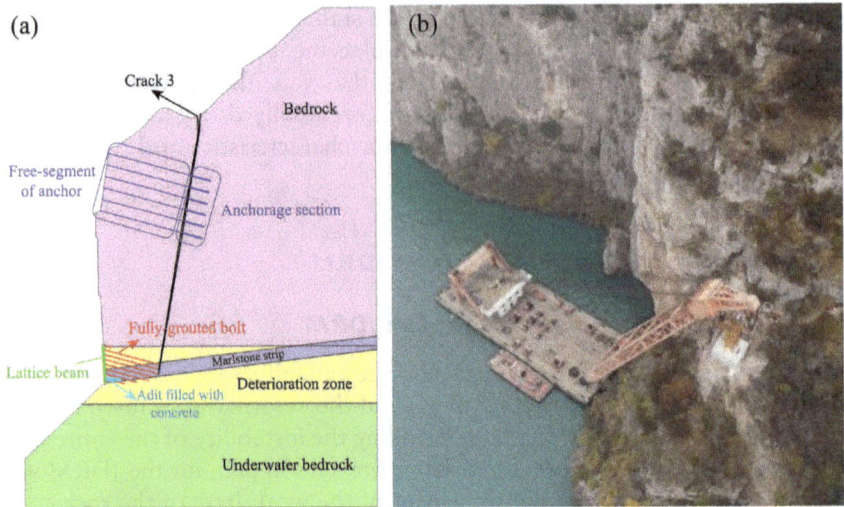

Figure 6.10 Preventative engineering techniques applied to the JDRM. (a) Side view illustration of the JDRM control measures. (b) Photograph of the on-site construction (January 2019).

Source: Wang et al. (2020b)

parameters obtained from the laboratory tests were reduced according to the standard of the Geo-hazard Prevention and Management Command of the Three Gorges Reservoir (2004). Based on the disaster prevention design proposed in the previous section, MIDAS software was used to carry out a finite element numerical calculation. The parameters used in the numerical calculation are shown in Table 6.7. In addition, the mechanical parameters of the protective materials were determined in accordance with the methods of Cai et al. (2006) and Li (2016). To study the effect of the various preventative methods on the stability of the dangerous rock mass, the reinforcement of the base of the rock mass and the anchoring of the upper part of the dangerous rock mass were independently numerically calculated. These two working conditions, together with the before protection and the comprehensive protection data, constitute the four working conditions calculated in this chapter.

In an attempt to study the effectiveness of the preventative methods, numeral calculations using the mechanical parameters for the initial state were carried out. Then, the displacement field and shear stress field of the JDRM under different working conditions were compared.

6.5.2.1 Influence of the preventative methods on the displacement field

The displacement field of the JDRM under different conditions is shown in Figure 6.11. This analysis showed that the displacement field of the JDRM

Table 6.7 Parameters used in the numerical calculation

Material category	Constitutive model	Elasticity modulus (MPa)	Poisson's ratio	Unit weight (kN/m³)	Cohesion (MPa)	Friction angle (°)	Elasticity modulus (MPa)
Underwater bedrock	Mohr-Coulomb	47800	0.26	27.2	5.21	44.4	47800
Degradation zone		42000	0.24	24.5	4.82	40.2	42000
Marlstone strip in degradation zone		2600	0.30	26.6	0.472	32.62	2600
Bedrock		50400	0.28	27.1	5.48	44.4	50400
Adit filled with concrete		28000	0.20	24.20	3.18	48.6	28000
Fully grouted bolt	Linear elasticity	196000	0.28	78.5	–	–	196000
Anchor		196000	0.28	78.5	–	–	196000
Lattice beam		27000	0.25	23.0	–	–	27000

Source: Wang et al. (2020b)

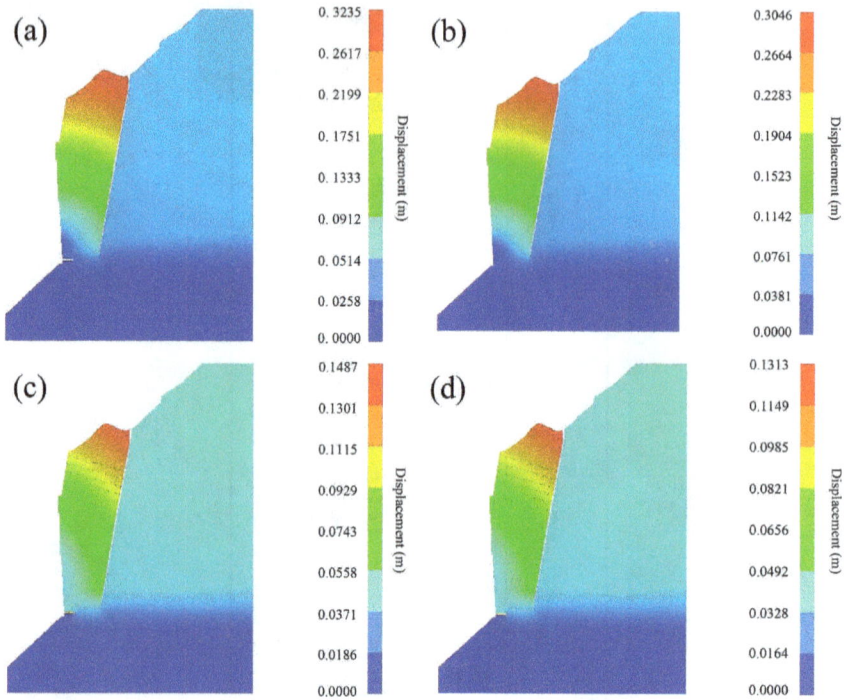

Figure 6.11 The displacement field of the JDRM under different working conditions: (a) before application of the preventative methods; (b) with reinforcement of the base of the rock mass; (c) with anchoring of the upper part of the rock mass; (d) with both preventative methods.

Source: Wang et al. (2020b)

was basically unchanged under the different working conditions. Before the preventative methods were applied, the maximum displacement of the dangerous rock mass was 0.3235 m. After both preventative methods were applied, the maximum displacement was 0.1313 m, which indicated that the displacement decreased by 59.38%. Thus, it can be concluded that preventative methods can effectively control the displacement of the dangerous rock mass. In addition, a comparison of the separate calculations of the various preventative methods indicated that anchoring the upper part of the dangerous rock mass was the key method in controlling the deformation of the dangerous rock mass.

6.5.2.2   Influence of preventative methods on the shear stress field

The shear strength field of the JDRM under different conditions is shown in Figure 6.12. Before the preventative methods were applied, the shear stress

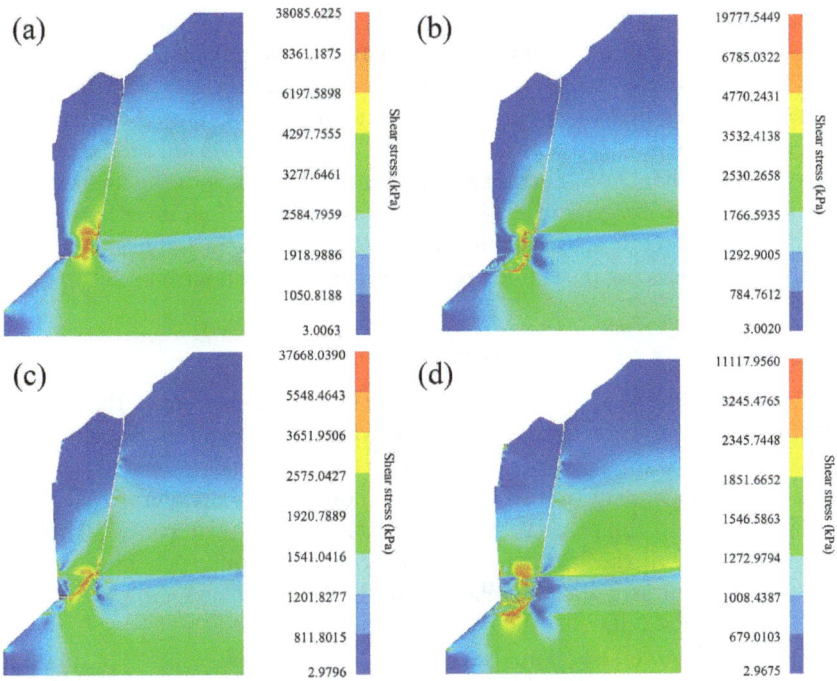

Figure 6.12 The shear stress field of the JDRM under different working conditions: (a) before application of the preventative methods; (b) with reinforcement of the base of the rock mass; (c) with anchoring of the upper part of the rock mass; (d) with both preventative methods.

Source: Wang et al. (2020b)

was concentrated at the intersection of the trailing edge crack and the degradation zone. After both of the preventative methods were applied, the maximum shear stress was 70.81% lower than that without the preventative methods, and the shear stress concentration zone was controlled by the reinforcement of the base of the rock mass. Thus, it can be concluded that preventative methods can effectively reduce the shear stress of the dangerous rock mass. In addition, by comparing the separate calculations of the various preventative methods, it can be found that the reinforcement of the base of the rock mass was the key method in controlling the concentration of the shear stress.

6.5.2.3 Influence of preventative methods on the FOS

The compressive strength of the base rock mass after the treatment was about 30 MPa (Jin and Zhao, 2002). The effective stress of the base rock mass in 2019 obtained by the prediction formula was 11.721 MPa. After substituting the effective stress into Equation (6.7), the FOS can be obtained, and it was 2.56. This FOS was 149 % higher than that of without treatment. In fact, in the process of treatment, the cracks of the base rock mass can be filled by the cement grouting, increasing the integrity of the base rock mass. Therefore, the effective stress will be lower than 11.721 MPa, and the FOS after the treatment will be higher than 2.56. Furthermore, with the increasing of the hydrological years, due to the integrity of the base rock mass, the fluctuation of water level had little effect on its strength, and the JDRM can be in a stable state for the long term. Besides, it was still necessary to carry out long-term monitoring.

6.6　Summary of this chapter

In this chapter, a new analysis method was proposed to study the damage evolution of the dangerous rock masses on the reservoir bank. This new method involved the generalization of a mechanical model, the derivation of the constitutive damage model based on laboratory results, iterative damage calculations, and the prediction of FOS based on the water level. Through this new method, the mechanical state of a dangerous rock mass can be tracked continuously and the relationship between the changing water level and the potential failure of the rock mass can be quantified.

By using this method to analyze JDRM, it can be found that the theoretical pressure data and the observed pressure data were consistent with one another, and both exhibited nonlinear deformation; this agreement showed that the new modeling technique was both valid and robust. As of December 2018, the calculations indicated that the JDRM will be in a critical state of instability after two dry–wet cycles.

According to the parameters obtained by the prediction model, the dynamic collapse analysis of JDRM was carried out. The collapse process of JDRM was similar to that of the Zengziyan collapse, which not only proved the effectiveness of parameter selection but also expanded the application scope.

In 2019, the treatment of JDRM was formally constructed. The simulation results indicated that anchoring the upper dangerous rock mass can effectively control the deformation of the JDRM, and reinforcement of the base rock mass can improve the shear strength of the JDRM. The combination of these two preventative methods can significantly improve the stability and satisfactorily ensure the long-term stability of the JDRM.

So far, the critical state judgment of JDRM was determined and the parameters were extracted to perform the dynamic collapse. Then, combining with the treatment of JDRM, the long-term stability of JDRM was predicted. The related calculation methods can provide significant references for the analysis of dangerous rock masses on the reservoir bank.

Chapter 7

Conclusions and outlook

7.1 Conclusions

Since water storage in the reservoir area in 2008, the fluctuation of reservoir water level has become an essential factor inducing the instability of the bank slope in the Three Gorges Reservoir area. According to the mechanical model of dangerous rock mass on the reservoir bank, a series of mechanical experiments with the help of improved test equipment that comprehensively considered multiple influencing factors were conducted. The damage evolution process of the steep dangerous rock masses on the reservoir bank was fully explored and concluded as follows:

(1) Establishment of mechanical model. According to the mechanical environment of Wuxia Gorge, the fluctuation of reservoir water level is determined as an essential factor inducing the instability of the bank slope, and the hazardous zone caused by the deterioration of the rocky bank continues to expand over time. For the mechanical environment of the dangerous rock mass on the reservoir bank, a comprehensive survey system was constructed to detect the development of cracks and broken areas in cliffs, underwater sections, and the interior of the dangerous rock mass. The filed survey, monitoring data, and indoor tests were used to generalize the mechanical model of JDRM. The self-weight of upper rock mass and the deterioration of the weak base rock mass caused by the reservoir water fluctuation are determined as the main factors leading to the deformation and failure of the JDRM.

(2) Establishment of the test environment. Through an improved experimental system, the samples can complete the conversion in the dry state and the hydraulic coupling state while the axial stress continued to act. Before the test, the polarizing microscope, ultra-high-resolution scanning electron microscope, and energy dispersive spectrum analysis were used to evaluate the microscopic mechanical properties of the selected samples qualitatively. Besides, the quality of the sample was controlled to reduce the dispersion of the test data through CT

DOI: 10.1201/9781003347163-7

scanning and wave speed testing. The results show that the selected sample has few pores and impurities, and these pores are difficult to form an active seepage channel.

(3) Analysis of hydraulic coupling test. Based on the actual mechanical state of the reservoir bank rock mass, the particle flow code, digital image correlation method, and acoustic emission technology were used to study the evolution process under different mechanical states. It can be found that the participation of water can reduce the bond contact and accelerate the deformation tendency between particles, which may eventually lead to an increase in plastic deformation and a decrease in peak strength. In the compression process under the hydraulic coupling state, since the water would not be transferred due to the compressed storage space, the water pressure continues to weaken the contact bond of the particles. Therefore, the hydraulic coupling state can accelerate the breaking of the contact bond between the particles and further promote the extension of the microcracks. And this is the reason for the difference in mechanical properties between the saturated state and the hydraulic coupling state. According to the entire evolution process of rock mass investigated by the test, it can be found that the damage can magnify the effective stress experienced by the base rock mass. Furthermore, incorporating the damage effects into the dangerous rock mass model creates a nonlinear and accelerating deformation trend that accurately portrays the failure of dangerous rock masses in the Three Gorges Reservoir area.

(4) Experimental research on the fluctuation of reservoir water level. According to the conventional dry–wet cycle test and numerical simulation, the difference in rock plasticity and peak strength under different mechanical conditions may become the key factor in accelerating the deterioration of rock mass. Through the transition of mechanical state tests under different axial pressures, it can be found that if the weight of the overlying rock mass were higher, the influence of mechanical transition on the base rock mass would be more significant. Compared with the creep experiments in a single state, the transition of the mechanical state continues to promote the degradation of the rock mass, dramatically reduces the peak strength of the rock mass, and accelerates the failure of the rock mass. Based on the field investigation and test results, the entire evolution of the dangerous rock mass on the reservoir bank can be divided into three stages: formation of the damage zone, progressive cumulative deformation, and nonlinear accelerated failure. Specifically, the periodic rise and fall of the water level in the reservoir area leads to the formation of a damage zone. Then, the periodic transition of the mechanical state further promotes cumulative deformation and reduces the peak intensity of the base of the rock mass, which causes the dangerous rock mass to be in a critical state faster. When the axial pressure exceeds the long-term strength, the evolution of this period nonlinearly accelerates.

(5) The evolution analysis and the treatment of JDRM. A new analysis method was proposed to study the damage evolution of the dangerous rock masses on the reservoir bank. This method involves the generalization of a mechanical model, the derivation of the constitutive damage model based on laboratory results, iterative damage calculations, and the prediction of FOS values based on the water level. Applying this method allows us to quantify the damage and the stability of the submerged dangerous rock mass over time. By using this method to analyze JDRM, it can be found that the theoretical pressure data and the observed pressure data are consistent with one another, and both exhibit nonlinear progressive deformation. As of December 2018, the results indicate that the JDRM will be in a critical state of instability after two dry–wet cycles. According to the parameters obtained by the prediction model, the dynamic collapse analysis of JDRM was carried out. The collapse process of JDRM is similar to that of the Zengziyan collapse, which proves the effectiveness of parameter selection. In 2019, the treatment of JDRM was formally constructed. The simulation results show that anchoring the upper dangerous rock mass can effectively control the deformation of the dangerous rock mass, and reinforcement of the base of the rock mass can improve the shear strength of the dangerous rock mass. These preventative methods can significantly improve the stability of the dangerous rock mass and satisfactorily ensure the long-term stability of the JDRM. The related calculation methods could provide significant references for the analysis of dangerous rock masses on the reservoir bank.

7.2   Outlook

In this book, the buckling failure of the steep dangerous rock masses on the reservoir bank is explored, indicating the damage evolution process caused by the reservoir water fluctuation. Multiple research methods are adopted in this research, including field investigation, in-situ monitoring, laboratory experiments, mathematical analysis, numerical simulation, etc. Since problems involved in reservoir banks are always complicated, and the time and current techniques are too limited to deal with all of the possible problems, several deficiencies and suggestions are summarized as follows for further research:

(1) The limestone is selected for experimental research in this book. Sandstone and other lithologies are also widely distributed in the Three Gorges Reservoir area, and related empirical studies can be analyzed using the same research ideas. The analysis method derived in this book is aimed at steep dangerous rock masses on the reservoir bank. Similar experimental methods and specific experiment environments can also be used for further analysis and research for the other rocky bank controlled by broken zones or structural planes.

(2) Due to the high density of limestone selected in this book, the seepage field is simplified. In the follow-up work, the seepage field of the rocky slope can be further studied, which may become a critical influencing factor for the rock masses that are more sensitive to water. Besides, although the relevant experimental conditions may be more complicated to set up, it is suggested to perform physical model experiments to better understand the evolution of the steep dangerous rock masses under self-weight.

(3) When carrying out the numerical calculation, the hydraulic coupling and the critical state are simplified. This approach could ensure the contrast of micro-parameters and the validity of numerical calculations. Since the calculation load of three-dimensional numerical simulation is relatively large, the two-dimensional numerical simulation is used for compression experiments and dynamic analysis in this book. In some cases, it is encouraged to further develop numerical calculation software from the perspective of the constitutive model and perform large-scale 3D numerical calculations by a workstation.

References

Abramson L W, Lee T S, Sharma S, et al. 2002. *Slope Stability and Stabilization Methods* (2nd ed.). New York: John Wiley & Sons.

Adaehi T, Oka F and Koike M. 2005. An elasto-viscoplastic constitutive model with strain-softening for soft sedimentary rocks. *Soils and Foundations*, 45(2): 125–133.

Adhikary D P and Dyskin A V. 2007. Modelling of progressive and instantaneous failure of foliated rock slopes. *Rock Mechanics and Rock Engineering*, 40(4): 349–362.

Arash B, Hwa K C and Tomoki S. 2014. Advanced structural health monitoring of concrete structures with the aid of acoustic emission. *Construction and Building Materials*, 66: 282–302.

Asadollahi P and Tonon F. 2012. Limiting equilibrium versus BS3D analysis of a tetrahedral rock block. *International Journal for Numerical and Analytical Methods in Geomechanics*, 36(1): 50–61.

Azzoni A, La Barbera G and Zaninetti A. 1995. Analysis and prediction of rockfalls using a mathematical model. *International Journal of Rock Mechanics and Mining Sciences and Geomechanics Abstracts*, 32(7): 709–724.

Bai M and Elsworth D. 1994. Modeling of subsidence and stress-dependent hydraulic conductivity for intact and fractured porous-media. *Rock Mechanics and Rock Engineering*, 27(4): 209–234.

Bargellini R, Halm D and Dragon A. 2006. Modelling of anisotropic damage by microcracks: towards a discrete approach. *Warszawa*, 58(1): 93–123.

Bendezu M, Romanel C and Roehl D. 2017. Finite element analysis of blast-induced fracture propagation in hard rocks. *Computers and Structures*, 182: 1–13.

Bi Y Z, He S M, Du Y J, et al. 2019. Effects of the configuration of a baffle-avalanche wall system on rock avalanches in Tibet Zhangmu: discrete element analysis. *Bulletin of Engineering Geology and the Environment*, 78(4): 2267–2282.

Bowman E T, Take W A, Rait K L, et al. 2012. Physical models of rock avalanche spreading behaviour with dynamic fragmentation. *Canadian Geotechnical Journal*, 49(4): 460–476.

Bozzano F, Martino S and Montagna A. 2012. A back analysis of a rock landslide to infer rheological parameters. *Engineering Geology*, 131–132: 45–56.

Braga M A S, Paquet H and Begonha A. 2002. Weathering of granites in a temperate climate (NW Portugal): granitic saprolites and arenization. *Catena*, 49(1–2): 41–56.

Cai M F, He M C and Liu D Y. 2006. *Rock Mechanics and Engineering*. China Science Press, Beijing.

Cao W G, Yuan J Z, Wang J Y, et al. 2013. A damage simulation technique of the full rock creep process considering accelerated creep. *Journal of Hunan University (Natural Sciences)*, 40(2): 15–20 (in Chinese).

Chang L H, Chen M Y, Jin W, et al. 2006. *Manual of Identification of Transparent Mineral Flakes*. Beijing: Geological Press.

Chen G Y and Lin Y M. 2004. Stress-strain-electrical resistance effects and associated state equations for uniaxial rock compression. *International Journal of Rock Mechanics and Mining Sciences*, 41(2): 223–236.

Chen X, He P and Qin Z. 2019. Strength weakening and energy mechanism of rocks subjected to wet-dry cycles. *Geotechnical and Geological Engineering*, 37(5): 3915–3923.

Chen X Y and Orense R P. 2020. Investigating the mechanisms of downslope motions of granular particles in small-scale experiments using magnetic tracking system. *Engineering Geology*, 265: 105448.

Chen Y F, Li D Q, Jiang Q H, et al. 2012. Micromechanical analysis of anisotropic damage and its influence on effective thermal conductivity in brittle rocks. *International Journal of Rock Mechanics and Mining Sciences*, 50: 102–116.

Cheng C, Li X, Li S D, et al. 2017. Failure behavior of granite affected by confinement and water pressure and its influence on the seepage behavior by laboratory experiments. *Materials*, 10(7): 798.

Cheng Q and Su S R. 2014. Movement characteristics of collapsed stones on slopes induced by Wenchuan earthquake. *Rock and Soil Mechanics*, 35(3): 772–776 (in Chinese).

Clough R W and Woodward R J. 1967. Analysis of embankment stresses and deformations. *Journal of Soil Mechanics and Foundations Division*, 93: 529–549.

Coombes M A and Naylor L A. 2012. Rock warming and drying under simulated intertidal conditions, part II: weathering and biological influences on evaporative cooling and near-surface micro-climatic conditions as an example of biogeomorphic ecosystem engineering. *Earth Surface Processes and Landforms*, 37(1): 100–118.

Cundall P A. 1971. A computer model for simulating progressive large scale movements in blocky rock systems. *Proceedings of the Symposium of the International Society for Rock Mechanics, Society for Rock Mechanics (ISRM)*, Nancy: France, II-8.

Cundall P A and Strack O D L. 1979. A discrete numerical model for granular assemblies. *Geotechnique*, 29(1): 47–65.

Daftaribesheli A, Ataei M and Sereshki F. 2011. Assessment of rock slope stability using the Fuzzy Slope Mass Rating (FSMR) system. *Applied Soft Computing*, 11(8): 4465–4473.

Dai Z W, Wang L Q, Zhang K Q, et al. 2022. Implementation of the Barton–Bandis nonlinear strength criterion into Mohr–Coulomb sliding failure model. *Advances in Materials Science and Engineering*, 1590884. https://doi.org/10.1155/2022/1590884

Damjanac B and Cundall P. 2016. Application of distinct element methods to simulation of hydraulic fracturing in naturally fractured reservoirs. *Computers and Geotechnics*, 71: 283–294.

Deng D P, Li L, Wang J F, et al. 2016. Limit equilibrium method for rock slope stability analysis by using the Generalized Hoek-Brown criterion. *International Journal of Rock Mechanics and Mining Sciences*, 89: 176–184.

Deng H F, Zhang Y C, Zhi Y Y, et al. 2019. Sandstone dynamical characteristics influenced by water-rock interaction of bank slope. *Advances in Civil Engineering*, 3279586.

Diederichs M S, Kaiser P K and Eberhardt E. 2004. Damage initiation and propagation in hard rock during tunneling and the influence of near-face stress rotation. *International Journal of Rock Mechanics and Mining Sciences*, 41: 785–812.

Doan M L and d'Hour V. 2012. Effect of initial damage on rock pulverization along faults. *Journal of Structural Geology*, 45: 111–122.

Dougill J W and Al E. 1976. Mechanics in Engineering. *Journal of Engineering Mechanics*, 333–355.

Dragon A and Mroz Z. 1979. A continuum model for plastic-brittle behaviour of rock and concrete. *International Journal of Engineering Science*, 17(2): 121–137.

Dussauge-Peisser C, Helmstetter A, Grasso J R, et al. 2002. Probabilistic approach to rock fall hazard assessment: potential of historical data analysis. *Natural Hazards and Earth System Sciences*, 2(1): 15–26.

Erarslan N and Williams D J. 2012a. The damage mechanism of rock fatigue and its relationship to the fracture toughness of rocks. *International Journal of Rock Mechanics and Mining Sciences*, 56: 15–26.

Erarslan N and Williams D J. 2012b. Mixed-mode fracturing of rocks under static and cyclic loading. *Rock Mechanics and Rock Engineering*, 46(5): 1035–1052.

Feng Z, Li B and He K. 2014a. Rock collapse mechanism on high-steep slope failure in sub-horizontal thick-bedded mountains. *Journal of Geomechanics*, 20: 123–131.

Feng Z, Li B, Yin Y P, et al. 2014b. Rockslides on limestone cliffs with subhorizontal bedding in the southwestern calcareous area of China. *Nature Hazards and Earth System Sciences*, 14: 2627–2635.

Feng Z, Yin Y P, Li B, et al. 2012. Centrifuge modeling of apparent dip slide from oblique thick bedding rock landslide. *Chinese Journal of Rock Mechanics and Engineering*, 31(5): 890–897 (in Chinese).

Ferrari F, Giacomini A and Thoeni K. 2016. Qualitative rockfall hazard assessment: a comprehensive review of current practices. *Rock Mechanics and Rock Engineering*, 49(7): 2865–2922.

Ferrero A M, Migliazza M R, Pirulli M, et al. 2016. Some open issues on rockfall hazard analysis in fractured rock mass: problems and prospects. *Rock Mechanics and Rock Engineering*, 49(9): 3615–3629.

Frayssines M and Hantz D. 2006. Failure mechanisms and triggering factors in calcareous cliffs of the subalpine ranges (French Alps). *Engineering Geology*, 86: 256–270.

Gao W. 2015. Stability analysis of rock slope based on an abstraction ant colony clustering algorithm. *Environmental Earth Sciences*, 73(12): 7969–7982.

Gao X J. 2013. Bifurcation behaviors of the two-state variable friction law of a rock mass system. *International Journal of Bifurcation and Chaos*, 23(11): 1350184.

Geo-hazard Prevention and Management Command of the Three Gorges Reservoirs. 2004. *Geological Survey Technical Handbook of Geo-hazard Prevention Project in the Three Gorges Reservoir* (in Chinese).

Ghazvinian E, Diederichs M S and Quey R. 2014. 3D random Voronoi grain-based models for simulation of brittle rock damage and fabric-guided micro-fracturing. *Journal of Rock Mechanics and Geotechnical Engineering*, 6: 506–521.

Griffiths D V and Lane P A. 1999. Slope stability analysis by finite elements. *Geotechnique*, 49(3): 387–403.

Griggs D T. 1939. Creep of rocks. *Journal of Geology*, 47: 225–251.

Guo W B, Hu B, Cheng J L, et al. 2020. Modeling time-dependent behavior of hard sandstone using the DEM method. *Geomechanics and Engineering*, 20(6): 517–525.

Haimson B C and Tharp T M. 1974. Stress around boreholes in bilinear elastic rock. *Society of Petroleum Engineers Journal*, 14(2): 145–151.

Hart R, Cundall P A and Lemos J. 1988. Formulation of a three-dimensional distinct element model—part II. Mechanical calculations for motion and interaction of a system composed of many polyhedral blocks. *International Journal of Rock Mechanics and Mining Sciences & Geomechanics Abstracts*, 25(3): 117–125.

Hawkes I, Mellor M and Gariepy S. 1973. Deformation of rocks under uniaxial tension. *International Journal of Rock Mechanics and Mining Sciences & Geomechanics Abstracts*, 10(6): 493–507.

He K. 2015. *Research on Collapse Mechanism of Tower Rock*. Xi'an: Chang'an University (in Chinese).

He K, Chen C L and Li B. 2019a. Case study of a rockfall in Chongqing, China: movement characteristics of the initial failure process of a tower-shaped rock mass. *Bulletin of Engineering Geology and the Environment*, 78(5): 3295–3303.

He K, Yin Y P, Li B, et al. 2019b. The mechanism of the bottom-crashing rockfall of a massive layered carbonate rock mass at Zengziyan, Chongqing, China. *Journal of Earth System Science*, 128: 104.

He Z P, Zhang Q Y, Wang J H, et al. 2007. Study on compressive creep test on diabasic dike at dam site of Dagangshan Hydropower Station. *Chinese Journal of Rock Mechanics and Engineering*, 26(12): 2495–2503 (in Chinese).

Herman G T. 2009. *Fundamentals of Computerized Tomography: Image Reconstruction from Projection* (2nd ed.). Cham: Springer International Publishing.

Hu H T and Zhao X Y. 2006. Studies on rockmass structure in slope of red bed in China. *Chinese Journal of Geotechnical Engineering*, 6: 689–694 (in Chinese).

Hu Y G, Lu W B, Wu X X, et al. 2018. Numerical and experimental investigation of blasting damage control of a high rock slope in a deep valley. *Engineering Geology*, 237: 12–20.

Hua W, Dong S M, Peng F, et al. 2017. Experimental investigation on the effect of wetting-drying cycles on mixed mode fracture toughness of sandstone. *International Journal of Rock Mechanics and Mining Sciences*, 93: 242–249.

Huang B L, Yin Y P and Du C L. 2016b. Risk management study on impulse waves generated by Hongyanzi landslide in Three Gorges Reservoir of China on June 24, 2015. *Landslides*, 13(3): 603–616.

Huang B L, Yin Y P, Liu G N, et al. 2012. Analysis of waves generated by Gongjiafang landslide in Wu Gorge, three Gorges reservoir, on November 23, 2008. *Landslides*, 9(3): 395–405.

Huang B L, Yin Y P, Wang S C, et al. 2014. A physical similarity model of an impulsive wave generated by Gongjiafang landslide in Three Gorges Reservoir, China. *Landslides*, 11(3): 513–525.

Huang B L, Zhang Z H, Yin Y P, et al. 2016a. A case study of pillar-shaped rock mass failure in the Three Gorges Reservoir Area, China. *Quarterly Journal of Engineering Geology and Hydrogeology*, 49: 195–202.

Huang R Q, Liu W H, Zhou J P, et al. 2007. Rolling tests on movement characteristics of rock blocks. *Chinese Journal of Geotechnical Engineering*, 9: 1296–1302 (in Chinese).

Hudson J A and Harrison J P. 2000. *Engineering Rock Mechanics: An Introduction to the Principles*. Burlington: Elsevier Science.

Hungr O and Evans S G. 2004. The occurrence and classification of massive rock slope failure. *Felsbau—Rock and Soil Engineering*, 22: 16–23.

Hungr O, Leroueil S and Picarelli L. 2014. The Varnes classification of landslide types, an update. *Landslides*, 11(2): 167–194.

Iverson R M, Reid M E and LaHusen R G. 1997. Debris-flow mobilization from landslides. *Annual Review of Earth and Planetary Sciences*, 25: 85–138.

Jeng F S, Lin M L and Huang T H. 2000. Wetting deterioration of soft sandstone-microscopic insights//ISRM international symposium. *International Society for Rock Mechanics*, 525.

Jia J Q, Liu X H, Zhang X, et al. 2013. Stability analysis of the wedge of rock slope and its programmed computation. *Journal of Computational and Theoretical Nanoscience*, 10(12): 2902–2905.

Jiang G L and Magnan J P. 1997. Stability analysis of embankments: comparison of limit analysis with methods of slices. *Geotechnique*, 47(4): 857–872.

Jiang Q H, Qi Y J, Wang Z J, et al. 2013. An extended Nishihara model for the description of three stages of sandstone creep. *Geophysical Journal International*, 193: 841–854.

Jin W L and Zhao Y X. 2002. State-of-the-art on durability of concrete structures. *Journal of Zhejiang University (Engineering Science)*, 4: 27–36 + 59 (in Chinese).

Kachanov L M. 1958. Time of the rupture process under creep conditions. *Nank SSR Otd Tech Nauk*, 8: 26–31.

Kachanov M L. 1982. A microcrack model of rock inelasticity part II: propagation of microcracks. *Mechanics of Materials*, 1(1): 29–41.

Kemeny J and Cook N G W. 1986. Effective moduli, non-linear deformation and strength of a cracked elastic solid. *International Journal of Rock Mechanics and Mining Sciences & Geomechanics Abstracts*, 23(2): 107–118.

Keneti A and Sainsbury B A. 2018. Review of published rockburst events and their contributing factors. *Engineering Geology*, 246: 361–373.

Kilburn C R J and Petley D N. 2003. Forecasting giant, catastrophic slope collapse: lessons from Vajont, Northern Italy. *Geomorphology*, 54(1–2): 21–32.

Krajcinovic D. 1984. Continuum damage mechanics. *Applied Mechanics Review*, 37(1): 15–20.

Krajcinovic D and Fonseka G U. 1981. The continuous damage theory of brittle materials. *Journal of Applied Mechanics*, 48(4): 809–815.

Kumar V, Gupta V and Jamir I. 2018. Hazard evaluation of progressive Pawari landslide zone, Satluj valley, Himachal Pradesh, India. *Natural Hazards*, 93(2): 1029–1047.

Lemaitre J. 1984. How to use damage mechanics. *Nuclear Engineering and Design*, 80(3): 233–245.

Li B, Xing A G and Xu C. 2017a. Simulation of a long-runout rock avalanche triggered by the Lushan earthquake in the Tangjia Valley, Tianquan, Sichuan, China. *Engineering Geology*, 218: 107–116.

Li G X. 2016. *Advanced in Soil Mechanics* (2nd ed.). Beijing: Tsinghua University Press (in Chinese).

Li L C, Tang C A, Li G, et al. 2012a. Numerical simulation of 3D hydraulic fracturing based on an improved flow-stress-damage model and a parallel FEM technique. *Rock Mechanics and Rock Engineering*, 45(5): 801–818.

Li L R, Deng J H, Zheng L, et al. 2017b. Dominant frequency characteristics of acoustic emissions in white marble during direct tensile tests. *Rock Mechanics and Rock Engineering*, 50(5): 1337–1346.

Li M. 2012. *Evaluation of High and Steep Slope Rock Mass Quality of Changbai Mountain Longmen Peak*. Changchun: Jilin University (in Chinese).

Li X, Cao W G and Su Y H. 2012b. A statistical damage constitutive model for softening behavior of rocks. *Engineering Geology*, 143–144: 1–17.

Li Z, Hu Z, Zhang X Y, et al. 2019. Reliability analysis of a rock slope based on plastic limit analysis theory with multiple failure modes. *Computers and Geotechnics*, 110: 132–147.

Lim K, Li A J, Schmid A, et al. 2017. Slope-stability assessments using finite-element limit-analysis methods. *International Journal of Geomechanics*, 17(2): 06016017.

Liu C Z. 2014. Genetic types of landslide and debris flow disasters in China. *Geological Review*, 60(4): 858–868 (in Chinese).

Liu D X and Cao P. 2015. Preliminary study of improved SMR method based on gray system theory. *Rock and Soil Mechanics*, 36: 408–412.

Liu S F, Lu S F, Wan Z J, et al. 2019. Investigation of the influence mechanism of rock damage on rock fragmentation and cutting performance by the discrete element method. *Royal Society Open Science*, 6(5): 190116.

Liu X R, Jin M H, Li D L, et al. 2018b. Strength deterioration of a Shaly sandstone under dry-wet cycles: a case study from the Three Gorges Reservoir in China. *Bulletin of Engineering Geology and the Environment*, 77(4): 1607–1621.

Liu X R, Wang Z J, Fu Y, et al. 2016. Macro/microtesting and damage and degradation of sandstones under dry-wet cycles. *Advances in Materials Science and Engineering*, 2016: 7013032.

Liu X, Wu Z J, Wang G, et al. 2013. Research on the softening and disintegration mechanism of carbonaceous shale. *Advanced Materials Research*, 671–674: 274–279.

Liu Z, Zhou C Y, Lu Y Q, et al. 2018a. Development of the multi-scale mechanical experimental system for rheological damage effect of soft rock bearing the hydro-mechanical coupling action. *Rock and Soil Mechanics*, 39(8): 3077–3086 (in Chinese).

Lux K H and Hou Z. 2000. New developments in mechanical safety analysis of repositories in rock salt. *Proceedings of the International Conference on Radioactive Waste Disposal, Disposal Technologies & Concepm*. Berlin: Springer Verlag, 281–286.

McDougall S and Hungr O. 2004. A model for the analysis of rapid landslide motion across three-dimensional terrain. *Canadian Geotechnical Journal*, 41(6): 1084–1097.

McMullan D. 2006. Scanning electron microscopy 1928–1965. *Scanning*, 17(3): 175–185.

Mitchell A and Hungr O. 2017. Theory and calibration of the Pierre 2 stochastic rock fall dynamics simulation program. *Canadian Geotechnical Journal*, 54(1): 18–30.

Molladavoodi H and Mortazavi A. 2011. A damage-based numerical analysis of brittle rocks failure mechanism. *Finite Elements in Analysis and Design*, 47: 991–1003.

Mortazavi A and Molladavoodi H. 2012. A numerical investigation of brittle rock damage model in deep underground openings. *Engineering Fracture Mechanics*, 90: 101–120.

Nedjar B and Le Roy R. 2013. An approach to the modeling of viscoelastic damage. Application to the long-term creep of gypsum rock materials. *International Journal for Numerical and Analytical Methods in Geomechanics*, 37(9): 1066–1078.

Nejati H R and Ghazvinian A. 2014. Brittleness effect on rock fatigue damage evolution. *Rock Mechanics and Rock Engineering*, 47(5): 1839–1848.

Neuman S P. 2005. Trends, prospects and challenges in quantifying flow and transport through fractured rocks. *Hydrogeology Journal*, 13(1): 124–147.

Nichol S L, Hungr O and Evans S G. 2002. Large-scale brittle and ductile toppling of rock slopes. *Canadian Geotechnical Journal*, 39(4): 773–788.

Nilsen B. 2017. Rock slope stability analysis according to Eurocode 7, discussion of some dilemmas with particular focus on limit equilibrium analysis. *Bulletin of Engineering Geology and the Environment*, 76(4): 1229–1236.

Nomikos P, Rahmannejad R and Sofianos A. 2011. Supported axisymmetric tunnels within linear viscoelastic burgers rocks. *Rock Mechanics and Rock Engineering*, 44(5): 553–564.

Özbek A. 2014. Investigation of the effects of wetting—drying and freezing—thawing cycles on some physical and mechanical properties of selected ignimbrites. *Bulletin of Engineering Geology and the Environment*, 73(2): 595–609.

Perkins T K and Krech W W. 1968. The energy balance concept of hydraulic fracturing. *Society of Petroleum Engineers Journal*, 8(1): 1–12.

Poisel R and Preh A. 2004. Rock slope initial failure mechanisms and their mechanical models. *Felsbau*, 22: 40–45.

Poisel R, Bednarik M, Holzer R, et al. 2005. Geomechanics of hazardous landslides. *Journal of Mountain Science*, 3: 211–217.

Poisel R, Hans A, Pollinger M, et al. 2009. Mechanics and velocity of the Larchberg-Galgenwald landslide (Austria). *Engineering Geology*, 109(1): 57–66.

Poisel R, Steger W and Zeitler A M P A. 1991. Stability investigations of competent rock masses lying on an incompetent base 7th ISRM congress. *International Society for Rock Mechanics*, 939–944.

Potyondy D O. 2015. The bonded-particle model as a tool for rock mechanics research and application: current trends and future directions. *Geosystem Engineering*, 18(1): 1–28.

Powell J W. 1875. *Exploration of the Colorado River of the West and Its Tributaries*. Washington, DC: Government Printing Office.

Qin Z C, Li T, Li Q H, et al. 2019. Mechanism of rock burst based on energy dissipation theory and its applications in erosion zone. *Acta Geodynamica et Geromaterialia*, 16(2): 119–130.

Reznichenko N V, Davies T R H and Alexander D J. 2011. Effects of rock avalanches on glacier behaviour and moraine formation. *Geomorphology*, 132(3–4): 327–338.

Rutqvist J and Stephansson O. 2003. The role of hydromechanical coupling in fractured rock engineering. *Hydrogeology Journal*, 11(1): 7–40.

Sarfaraz H and Amini M. 2020. Numerical modeling of rock slopes with a potential of block-flexural toppling failure. *Journal of Mining and Environment*, 11(1): 247–259.

Scaringi G, Fan X M, Xu Q, et al. 2018. Some considerations on the use of numerical methods to simulate past landslides and possible new failures: the case of the recent Xinmo landslide (Sichuan, China). *Landslides*, 15(7): 1359–1375.

Shao J F, Chau K T and Feng X T. 2006. Modeling of anisotropic damage and creep deformation in brittle rocks. *International Journal of Rock Mechanics and Mining Sciences*, 43: 582–592.

Shao P, Xu Z W, Zhang H Q, et al. 2009. Evolution of blast-induced rock damage and fragmentation prediction. *Procedia Earth and Planetary Science*, 1: 585–591.

Shen X H, Wei Y J, Tao L J, et al. 2012. Mathematical models and application of unstable rock masses collapsing movement along slopes. *Journal of Beijing University of Technology*, 38(4): 540–543 (in Chinese).

Shi G H. 1988. *Discontinuous Deformation Analysis: A New Numerical Model for the Statics and Dynamics of Block Systems*. Berkeley, CA: University of California, Berkeley.

Sonnekus M and Smith J V. 2018. A review of selected unexpected large slope failures. *International Journal of Geomate*, 15(48): 66–73.

Sosio R, Crosta G B and Hungr O. 2008. Complete dynamic modeling calibration for the Thurwieser rock avalanche (Italian Central Alps). *Engineering Geology*, 100(1–2): 11–26.

Specification for rock tests in water conservancy and hydroelectric engineering. 2001. Industry Standards of the People's Republic of China. SL264–2001.

Standard for test methods of engineering rock mass. 2013. National Standards of the People's Republic of China. GBT50266–2013.

Stead D, Eberhardt E and Coggan J S. 2006. Developments in the characterization of complex rock slope deformation and failure using numerical modelling techniques. *Engineering Geology*, 83(1–3): 217–235.

Stead D and Wolter A. 2015. A critical review of rock slope failure mechanisms: the importance of structural geology. *Journal of Structural Geology*, 74: 1–23.

Stimpson B and Chen R. 1993. Measurement of rock elastic moduli in tension and in compression and its practical significance. *Canadian Geotechnical Journal*, 30(2): 338–347.

Sumner P D and Loubser M J. 2008. Experimental sandstone weathering using different wetting and drying moisture amplitudes. *Earth Surface Processes and Landforms*, 33(6): 985–990.

Sun S R, Xu P L, Wu J M, et al. 2014. Strength parameter identification and application of soil-rock mixture for steep-walled talus slopes in southwestern China. *Bulletin of Engineering Geology and the Environment*, 73(1): 123–140.

Tang C A. 1997. Numerical simulation of progressive rock failure and associated seismicity. *International Journal of Rock Mechanics and Mining Sciences*, 34(2): 249–261.

Tang H D and Zhu M L. 2020. Micro damage mechanics-based exponential power law acceleration of microscale damage and time-dependent crack propagation. *Engineering Fracture Mechanics*, 229: 106930.

Tang H M, Wasowski J and Juang C H. 2019. Geohazards in the three Gorges Reservoir Area, China—lessons learned from decades of research. *Engineering Geology*, 261: 105267.

Terzaghi K. 1950. Mechanism of landslides. *The Geological Society of America, Engineering Geology (Berkey)*, 83–123.

Thompson N, Bennett M R and Petford N. 2009. Analyses on granular mass movement mechanics and deformation with distinct element numerical modeling: implications for large-scale rock and debris avalanches. *Acta Geotechnica*, 4(4): 233–247.

Tomanovic Z. 2006. Rheological model of soft rock creep based on the tests on marl. *Mechanics of Time-dependent Materials*, 10(2): 135–154.

Tomas R, Cuenca A, Cano M, et al. 2012. A graphical approach for slope mass rating (SMR). *Engineering Geology*, 124: 67–76.

Tonon F. 2020. Simplified consideration for permanent rock dowels in block theory and 2-D limit equilibrium analyses. *Rock Mechanics and Rock Engineering*, 53(4): 2001–2006.

Torres-Suarez M C, Alarcon-Guzman A and Berdugo-De Moya R. 2014. Effects of loading—unloading and wetting—drying cycles on geomechanical behaviors of mudrocks in the Colombian Andes. *Journal of Rock Mechanics and Geotechnical Engineering*, 6(3): 257–268.

Ulusay R. 2014. *The ISRM Suggested Methods for Rock Characterization, Testing and Monitoring: 2007–2014*. Cham: Springer International Publishing.

Valanis K C. 1971. A theory of visco-plasticity without a yield surface. *Archives of Mechanics*, 23(5): 517–55l.

Wang D P, Wang Z W, Li Y Z, et al. 2020a. Characteristics and dynamic process analysis of the 2018. Mabian consequent landslide in Sichuan Province, China. *Bulletin of Engineering Geology and the Environment*, 79: 3337–3359.

Wang H H, Han S H and Zhou X Y. 1991. Experiment study of dangerous rock mass at Lianziya dangerous rock mass cliff in the Thress Gorges by centrifuge. *Construction and Design Research*, 4: 27–31 (in Chinese).

Wang L Q, Yin Y P, Huang B L, et al. 2020b. A study of the treatment of a dangerous thick submerged rock mass in the Three Gorges Reservoir Area. *Bulletin of Engineering Geology and the Environment*, 79: 2579–2590.

Wang L Q, Yin Y P, Huang B L, et al. 2020d. Damage evolution and stability analysis of the Jianchuandong Dangerous Rock Mass in the Three Gorges Reservoir Area. *Engineering Geology*, 265: 105439.

Wang L Q, Yin Y P, Zhou C Y, et al. 2020c. Damage evolution of hydraulically coupled Jianchuandong dangerous rock mass. *Landslides*, 17: 1083–1090.

Wang W P, Li B, Huang B L, et al. 2016b. Stability analysis of sub-horizontal thickbedded slope in the Three Gorges reservoir area: a case study of Jianchundong dangerous rock mass in Wushan, Chongqing. *Journal of Geomechanics*, 22(3): 725–732 (in Chinese).

Wang X G. 2014. *Study on the Rheological Mechanics and Engineering Application of Rocks of High Reservoir Bank Slope under the Deterioration Effect of Water Saturation-Dehydration Circulation*. Wuhan: China University of Geosciences (in Chinese).

Wang X G, Hu B, Hu X L, et al. 2016c. A constitutive model of granite shear creep under moisture. *Journal of Earth Science*, 27(4): 677–685.

Wang X G, Wang J D, Gu T F, et al. 2017. A modified Hoek-Brown failure criterion considering the damage to reservoir bank slope rocks under water saturation-dehydration circulation. *Journal of Mountain Science*, 14(4): 771–781.

Wang X G, Yin Y P, Wang J D, et al. 2018. A nonstationary parameter model for the sandstone creep tests. *Landslides*, 15: 1377–1389.

Wang Y F, Xu Q, Cheng Q G, et al. 2016a. Experimental study on the propagation and deposit features of rock avalanche along 3D complex topography. *Chinese Journal of Rock Mechanics and Engineering*, 35(9): 1776–1791 (in Chinese).

Wang Z J. 2016. *Damage Evolution Characteristics and the Accumulation Damage Model of Sandstone Under Dry-Wet Cycle*. Chongqing: Chongqing University (in Chinese).

Wang Z L, Li Y C and Wang J G. 2008. A method for evaluating dynamic tensile damage of rock. *Engineering Fracture Mechanics*, 75: 2812–2825.

Wei J, Zhu W C, Guan K, et al. 2020. An acoustic emission data-driven model to simulate rock failure process. *Rock Mechanics and Rock Engineering*, 53(4): 1605–1621.

Wu F Q, Deng Y, Wu J, et al. 2020. Stress-strain relationship in elastic stage of fractured rock mass. *Engineering Geology*, 268: 105498.

Wu L Z, Li B, Huang R Q, et al. 2017. Experimental study and modeling of shear rheology in sandstone with non-persistent joints. *Engineering Geology*, 222: 201–211.

Xia K Z, Liu X M, Chen C X, et al. 2015. Analysis of mechanism of bedding rock slope instability with catastrophe theory. *Rock and Soil Mechanics*, 32(2): 477–486.

Xing A G, Wang G H, Yin Y P, et al. 2016. Investigation and dynamic analysis of a catastrophic rock avalanche on September 23, 1991, Zhaotong, China. *Landslides*, 13(5): 1035–1047.

Xing A G, Yuan X Y, Xu Q, et al. 2017. Characteristics and numerical runout modelling of a catastrophic rock avalanche triggered by the Wenchuan earthquake in the Wenjia valley, Mianzhu, Sichuan, China. *Landslides*, 14(1): 83–98.

Xu W H, Kang Y F, Chen L C, et al. 2022. Dynamic assessment of slope stability based on multi-source monitoring data and ensemble learning approaches: a case study of Jiuxianping landslide. *Geological Journal*. https://doi.org/10.1002/gj.4605

Xu W Y, Wang R B, Wang W, et al. 2012. Creep properties and permeability evolution in triaxial rheological tests of hard rock in dam foundation. *Journal of Central South University of Technology*, 19(1): 252–261.

Xu X J, Liu B, Li S C, et al. 2016. The electrical resistivity and acoustic emission response law and damage evolution of limestone in Brazilian split test. *Advances in Materials Science and Engineering*, 8052972.

Xue J J and Zhang Z H. 2011. Experimental research on relationship between strength of sandstone and wave velocity during wet and dry cycle. *Journal of China Three Gorges University (Natural Sciences)*, 33(3): 51–54 (in Chinese).

Ya N, Wang L S, Zhao Q H, et al. 1996. Simulation study of rockfall kinematics. *Journal of Geological Hazards and Environment Preservation*, 2: 25–32 (in Chinese).

Yan C Z, Zheng H, Sun G H, et al. 2016. Combined finite-discrete element method for simulation of hydraulic fracturing. *Rock Mechanics and Rock Engineering*, 49(4): 1389–1410.

Yang B, Mo S H, Wu P, et al. 2013a. An empirical constitutive correlation for regular jugged discontinuity of rock surfaces. *Advances in Applied Mathematics and Mechanics*, 5(2): 258–268.

Yang H W. 2011. *Study on Coupling Mechanism of Rock and Pore Water under Cyclic Loading*. Chongqing: Chongqing University (in Chinese).

Yang T H, Tham L G, Tang C A, et al. 2004. Influence of heterogeneity of mechanical properties on hydraulic fracturing in permeable rocks. *Rock Mechanics and Rock Engineering*, 37(4): 251–275.

Yang T H, Xu T, Liu H Y, et al. 2013c. Rheological characteristics of weak rock mass and effects on the long-term stability of slopes. *Rock Mechanics and Rock Engineering*, 47(6): 2253–2263.

Yang W D, Zhang Q Y, Li S C, et al. 2013b. Time-dependent behavior of diabase and a nonlinear creep model. *Rock Mechanics and Rock Engineering*, 47: 1211–1224.

Yao H Y, Zhang Z H, Zhu Z H, et al. 2010. Experimental study of mechanical properties of sandstone under cyclic drying and wetting. *Rock and Soil Mechanics*, 31(12): 3704–3708.

Yao H Y, Zhu Y and Wu P. 2013. Research on uniaxial compression and tension tests of sandstone subjected to drying-wetting cycle. *Disaster Advances*, 6: 388–392.

Yin Y P, Huang B L, Liu G N, et al. 2015. Potential risk analysis on a Jianchuandong dangerous rockmass-generated impulse wave in the Three Gorges Reservoir, China. *Environmental Earth Sciences*, 2595–2607.

Yin Y P, Huang B L, Wang W P, et al. 2016. Reservoir-induced landslides and risk control in Three Gorges Project on Yangtze River, China. *Journal of Rock Mechanics and Geotechnical Engineering*, 577–595.

Yin Y P, Li B, Gao Y, et al. 2023. Geostructures, dynamics and risk mitigation of high-altitude and long-runout rockslides. *Journal of Rock Mechanics and Geotechnical Engineering*. https://doi.org/10.1016/j.jrmge.2022.11.001

Yin Y P, Sun P, Zhu J L, et al. 2011. Research on catastrophic rock avalanche at Guanling, Guizhou, China. *Landslides*, 8(4): 517–525.

Yin Y P, Wang L Q, Dai Z W, et al. 2022. Evolution analysis of the Banbiyan Dangerous Rock Mass in the Three Gorges Reservoir area, China. *Georisk: Assessment and Management of Risk for Engineered Systems and Geohazards NGRK*, https://doi.org/10.1080/17499518.2022.2062776

Yu Y J, Zou C N, Dong D Z, et al. 2014. Geological conditions and prospect forecast of shale gas formation in Qiangtang Basin, Qinghai-Tibet Plateau. *Acta Geologica Sinica-English Edition*, 88(2): 598–619.

Yuan H P, Cao P, Xu W Z, et al. 2006. Visco-elato-plastic constitutive relationship of rock and modified Burgers creep model. *Chinese Journal of Geotechnical Engineering*, 28(6): 796–799 (in Chinese).

Yuan S C and Harrison J P. 2006. A review of the state of the art in modelling progressive mechanical breakdown and associated fluid flow in intact heterogeneous rocks. *International Journal of Rock Mechanics and Mining Sciences*, 43: 1001–1022.

Zeng Z X, Kong L W, Tian H, et al. 2017. Effect of drying and wetting cycles on disintegration behavior of swelling mudstone and its grading entropy characterization. *Rock and Soil Mechanics*, 38(7): 1983–1989.

Zhang B Y, Zhang J H and Sun G L. 2015. Deformation and shear strength of rockfill materials composed of soft siltstones subjected to stress, cyclical drying/wetting and temperature variations. *Engineering Geology*, 190: 87–97.

Zhang C, Cai Z Y, Huang Y H, et al. 2018. Laboratory and centrifuge model tests on influence of swelling rock with drying-wetting cycles on stability of canal slope. *Advances in Civil Engineering*, 4785960.

Zhang J, Standifird W B, Roegiers J C, et al. 2007b. Stress-dependent fluid flow and permeability in fractured media: from lab experiments to engineering applications. *Rock Mechanics and Rock Engineering*, 40(1): 3–21.

Zhang K Q, Wang L Q, Dai Z W, et al. 2022b. Evolution trend of the Huangyanwo rock mass under the action of reservoir water fluctuation. *Natural Hazards*, 113: 1583–1600.

Zhang K Q, Wang L Q, Wengang Z, et al. 2021. Formation and failure mechanism of the Xinfangzi landslide in Chongqing City (China). *Applied Sciences-Basel*, 11(19): 8963.

Zhang W G, Li H R, Tang L B, et al. 2022d. Displacement prediction of Jiuxian-ping landslide using gated recurrent unit (GRU) networks. *Acta Geotechnica*, 17: 1367–1382.

Zhang W G, Liu S L, Wang L Q, et al. 2022c. Landslide susceptibility research combining qualitative analysis and quantitative evaluation: a case study of Yunyang County in Chongqing, China. *Forests*, 13(7): 1055.

Zhang W G, Meng X Y, Wang L Q, et al. 2022a. Stability analysis of the reservoir bank landslide with weak interlayer considering the influence of multiple factors. *Geomatics, Natural Hazards & Risk*, 13(1): 2911–2924.

Zhang X K, Qin S Q, Li Z G, et al. 2007a. Stability analysis of an unstable rock block at the site of Xilongchi pumped storage power station. *Journal of Engineering Geology*, 2: 174–178.

Zhao D, Swoboda G and Laabmayr F. 2004. Damage mechanics and its application for the design of an underground theater. *Tunnelling and Underground Space Technology*, 19: 567–575.

Zhao J S, Feng X T, Jiang Q, et al. 2018. Microseismicity monitoring and failure mechanism analysis of rock masses with weak interlayer zone in underground intersecting chambers: a case study from the Baihetan Hydropower Station, China. *Engineering Geology*, 245: 44–60.

Zhao Y L, Wang Y X, Wang W J, et al. 2017. Modeling of non-linear rheological behavior of hard rock using triaxial rheological experiment. *International Journal of Rock Mechanics and Mining Sciences*, 93: 66–75.

Zhao Z H, Jing L R and Neretnieks I. 2012. Particle mechanics model for the effects of shear on solute retardation coefficient in rock fractures. *International Journal of Rock Mechanics and Mining Sciences*, 52: 92–102.

Zheng H H, Li T B, Shen J Y, et al. 2018. The effects of blast damage zone thickness on rock slope stability. *Engineering Geology*, 246: 19–27.

Zhou C Y, Lu Y Q, Liu Z, et al. 2019. An innovative acousto-optic-sensing-based triaxial testing system for rocks. *Rock Mechanics and Rock Engineering*, 52(9): 3305–3321.

Zhou C Y and Zhu F X. 2010. An elasto-plastic damage constitutive model with double yield surfaces for saturated soft rock. *International Journal of Rock Mechanics and Mining Sciences*, 47: 385–395.

Zhou H W, Wang C P and Han B. 2011. A creep constitutive model for salt rock based on fractional derivatives. *International Journal of Rock Mechanics and Mining Sciences*, 48(1): 116–121.

Zhou J W, Xu W Y and Yang X G. 2010. A microcrack damage model for brittle rocks under uniaxial compression. *Mechanics Research Communications*, 37(4): 399–405.

Zhou X, Chen J P, Chen Y, et al. 2017a. Bayesian-based probabilistic kinematic analysis of discontinuity-controlled rock slope instabilities. *Bulletin of Engineering Geology and the Environment*, 76(4): 1249–1262.

Zhou X P, Lian Y J, Wong L N Y, et al. 2018. Understanding the fracture behavior of brittle and ductile multi-flawed rocks by uniaxial loading by digital image correlation. *Engineering Fracture Mechanics*, 199: 438–460.

Zhou Z, Cai X, Chen L, et al. 2017b. Influence of cyclic wetting and drying on physical and dynamic compressive properties of sandstone. *Engineering Geology*, 220: 1–12.

Zhou Z L, Cai X, Cao W Z, et al. 2016. Influence of water content on mechanical properties of rock in both saturation and drying processes. *Rock Mechanics and Rock Engineering*, 49(8): 3009–3025.

Zhu C H, Wu P, Yao H Y, et al. 2012. Split test of sandstone under condition of cyclic saturation-drying and long-term saturation. *Water Resources and Power*, 30(12): 58–60 (in Chinses).

Zhu D Y, Lee C F and Jiang H D. 2003. Generalized framework of limit equilibrium methods for slope stability analysis. *Geotechnique*, 53 (4): 377–395.

Zou Y L, Zhang W G, Wang L Q, et al. 2022. Cross-scale study of the high-steep reservoir banks under different mechanical states. *Lithosphere*, 1: 1077678.

Zuo Q H, Disilvestro D and Richter J D. 2010. A crack-mechanics based model for damage and plasticity of brittle materials under dynamic loading. *International Journal of Solids and Structures*, 47: 2790–2798.

Index

Note: Page numbers in *italics* indicate a figure and page numbers in **bold** indicate a table on the corresponding page.

For Product Safety Concerns and Information please contact our EU
representative GPSR@taylorandfrancis.com
Taylor & Francis Verlag GmbH, Kaufingerstraße 24, 80331 München, Germany